Aerial Life

RGS-IBG Book Series

Published

Forthcoming

Aerial Life

Spaces, Mobilities, Affects

Peter Adey

WILEY-BLACKWELL

A John Wiley & Sons, Ltd., Publication

This edition first published 2010
© 2010 Peter Adey

Blackwell Publishing was acquired by John Wiley & Sons in February 2007. Blackwell's publishing program has been merged with Wiley's global Scientific, Technical, and Medical business to form Wiley-Blackwell.

Registered Office
John Wiley & Sons Ltd, The Atrium, Southern Gate, Chichester, West Sussex, PO19 8SQ, United Kingdom

Editorial Offices
350 Main Street, Malden, MA 02148-5020, USA
9600 Garsington Road, Oxford, OX4 2DQ, UK
The Atrium, Southern Gate, Chichester, West Sussex, PO19 8SQ, UK

For details of our global editorial offices, for customer services, and for information about how to apply for permission to reuse the copyright material in this book please see our website at www.wiley.com/wiley-blackwell.

The right of Peter Adey to be identified as the author of this work has been asserted in accordance with the UK Copyright, Designs and Patents Act 1988.

Wiley also publishes its books in a variety of electronic formats. Some content that appears in print may not be available in electronic books.

Designations used by companies to distinguish their products are often claimed as trademarks. All brand names and product names used in this book are trade names, service marks, trademarks or registered trademarks of their respective owners. The publisher is not associated with any product or vendor mentioned in this book. This publication is designed to provide accurate and authoritative information in regard to the subject matter covered. It is sold on the understanding that the publisher is not engaged in rendering professional services. If professional advice or other expert assistance is required, the services of a competent professional should be sought.

Library of Congress Cataloging-in-Publication Data

Adey, Peter.
  Aerial life : spaces, mobilities, affects / Peter Adey.
    p.   cm. – (RGS-IBG book series)
  Includes bibliographical references and index.
  ISBN 978-1-4051-8262-1 (hardcover : alk. paper) – ISBN 978-1-4051-8261-4 (pbk. : alk. paper)
1. Aeronautics–Social aspects.   2. Air travel–Social aspects.   3. Human geography.   4. Social mobility.   I. Title.
  TL553.A34 2010
  306.4'819–dc22

                                                        2009052085

A catalogue record for this book is available from the British Library.

Set in 10/12pt Plantin by SPi Publisher Services, Pondicherry, India
Printed and bound in Malaysia by Vivar Printing Sdn Bhd

I   2010

For us

Contents

Figures and Tables

Series Editors' Preface

The RGS-IBG Book Series only publishes work of the highest international standing. Its emphasis is on distinctive new developments in human and physical geography, although it is also open to contributions from cognate disciplines whose interests overlap with those of geographers. The Series places strong emphasis on theoretically informed and empirically strong texts. Reflecting the vibrant and diverse theoretical and empirical agendas that characterize the contemporary discipline, contributions are expected to inform, challenge and stimulate the reader. Overall, the RGS-IBG Book Series seeks to promote scholarly publications that leave an intellectual mark and change the way readers think about particular issues, methods or theories.

For details on how to submit a proposal please visit:
www.rgsbookseries.com

Kevin Ward
University of Manchester, UK

Joanna Bullard
Loughborough University, UK

RGS-IBG Book Series Editors

Acknowledgements

This book has been in the making for almost 10 years. Since a 2nd year geography fieldtrip to New York in 2000 left me contemplating Heathrow's strange terminal landscape, I've been hooked on all things aerial. This was followed through in postgraduate work and a post-doctoral fellowship under the supervision and mentorship of Tim Cresswell, Deborah Dixon and Martin Jones, all of whom failed to suffer my writing in silence!

Thanks really must go to Kevin Ward and Jacqueline Scott as the book has sailed through the publishing process. I also appreciate the academic and creative freedom that the RGS-IBG series at Wiley-Blackwell encourages.

Conversations and collaboration with colleagues at Keele such as Luis Lobo Guerrero altered my focus and helped broaden my reading and thinking. Zoe Robinson and Peter Knight gave a different perspective on an initial proposal draft, and Barry Godfrey gave a lot of support and helped work through some of the ideas and material on bombing and the Blitz. Other *planeurs* have also been a constant source of criticism, inspiration and support; these include, particularly: Ben Anderson, Jon Anderson, David Bissell, Lucy Budd, Rachel Colls, Dave Cox, Gillian Fuller, Steve Graham, John Horton, Ole B. Jensen, Pete Kraftl, Lisa Lau, Deirdre McKay, Craig Martin, Pete Merriman, Steve Quilley, Mark B. Salter, John Urry and Chris Zebrowski. James Sidaway also kindly sent me his startling image of the networks of extraordinary rendition.

Along the way, various aspects of the book were presented and discussed at many conferences and especially departmental seminars, where discussion added so much to its tone and focus. University venues here included Durham, Bristol, Edinburgh, Manchester and, lastly, UCL, who permitted me a semester of office space and support thanks to James Kneale. Source material came from consultation of numerous archival collections, including Birmingham, Cardiff and Wolverhampton City Archives; the Modern Records Centre, Warwick; the Hagley Museum and Library; the National

Archives; the Imperial War Museum; the British Library; the RAF Museum; the Air League Archives; and the Scout Association Library, whose archivist, Pat Styles, was particularly helpful. I also acknowledge thanks to the ICAO and the Royal Society of the Arts for kind permission to reproduce several figures, and especially the Friends of Liverpool Airport to use the photos from the Alan Thelwell Collection. The book would never have been born let alone reached completion without the generous support from an ESRC PhD studentship, ESRC post-doctoral fellowship and finally an AHRC research leave award scheme funding which gave me a year's worth of writing time.

Aerial Life has been a group effort. Conversations with Grandad and Nanna Searles about the RAF and the war added colour to my documentary efforts. But it was also written sometimes in difficult health and circumstances. Thanks to my wife and our family for their love in helping us see this project through.

Every effort has been made to trace copyright holders and to obtain their permission for the use of copyright material. The publisher apologizes for any errors or omissions in the above list and would be grateful if notified of any corrections that should be incorporated in future reprints or editions of this book.

Chapter One

Introduction

Prologue

Responding to the rapid growth in commercial air transportation and its implications for the personal mobilities and geopolitical relations that follow, French urban sociologist Henri Lefebvre would write that 'space is also being recast' (1991: 351). This book is concerned with the scale and content of that recasting.

1920, the Royal Geographical Society, Kensington Gore, London

A meeting held at the Royal Geographical Society in Kensington Gore, London, brought together geographers and eminent figures who shared interests in the advancement of British military and civilian aviation. Winston Churchill, Hugh Trenchard, Frederick Sykes and Geoffrey Salmond[1] included, the group set about discussing the major milestones passed by a growing British air network and the creeping establishment of new routes in Africa and the colonial territories. Opening the meeting, the Society's President, Sir Francis Younghusband, spoke about the relation between the field of geography and the burgeoning technology of the aeroplane.

> We earth-bound geographers are inclined to look with a jealous eye upon these fine gentlemen of the air. For they soar up aloft and glide gracefully over the most terrible obstacles, insurmountable to us geographers. We dislike them especially for a very nasty habit they have contracted of taking photographs of us from that superior position in which men appear like ants, mountains like mole-hills, and even the President of the Royal Geographical Society

appears of very insignificant proportions. But we geographers get our own back upon them in the long run, because they cannot stay up in the air for ever. (Prince of Wales et al. 1920: 263)

As an example of coming to terms with the new sorts of perspectives available from the aeroplane, the meeting was typical. The aeroplane was to develop quite new ways of seeing space and time.[2] From the aircraft the earth tilted. It became a large canvas as the embodied gaze left the shackles of the terrestrial subject. Quite different people would be necessary to take advantage of these viewpoints and unfettered, frictionless movements. Together, entirely new sorts of space would be born.

On the other hand, the geographers diverged from what are now rather typical conceptions of mobility or air travel in favour of a position that concerned the ground the aeroplane was dependent upon. For Younghusband, it was questions of geography – questions concerning the surface of the planet that earth-bound the airmen.

If they are in an aeroplane they are most anxious that the surface of the earth beneath them is not water, and if they are in a flying-boat they do not want it to be land. [...] They want to know if it is covered with forests or buildings, whether it is hilly or plains, whether it is crowded or free and open, and whether there are communications to their landing-place. They want, in fact, to know everything they can about its geography. (Prince of Wales et al. 1920: 263)

In the end, Younghusband joked, they would be 'glad enough, these haughty airmen, to shake hands with us humble geographers' (Prince of Wales et al. 1920: 263).

At first Younghusband appears to separate the glamorous lives of the airmen – who act on high above from their privileged perspectives – from the existence of the geographer – whose concern is with the terrestrial: with the landforms, patternings and peoples of the ground below. These distinctions are commonplace 'as separations between above and below, air and ground, bomber and bombed' are made in representation (Gregory 2010 in press). At the same time, however, the world of the airman is made dependent on the world underneath. New connections between the horizontal and the vertical are brought into being. Like the airmen and the geographer, the two dimensions shake hands. The vertical life of the airmen is tethered to the terrestrial, to the location of their landing place and its local topographical idiosyncrasies, as well as its outer connectivities. The aeroplane is not necessarily liberating or liberated, but it is tied, or what John Urry (2003) might describe as 'moored', to an infrastructure on the ground. Both the ground and the air reside together in vertical reciprocity.

1941, Dachau concentration camp, Germany

In *Homo Sacer*, Giorgio Agamben (1998) discusses the work of German Nazi scientist Dr Sigmund Rascher, who in 1941 applied to Himmler to be provided with two or three 'professional criminals' for the purposes of his research on the pilot body. The research was commissioned for the Luftwaffe. His experiments required VPs, short for *Versuchpersonen* (human guinea pigs), who could be stretched to the limits and beyond of the human body. His experiments would test their abilities to withstand, adapt to and eventually succumb to the environments of high-altitude flying, or submersion in freezing temperatures – to simulate the conditions of their pilot counterparts should they crash-land or parachute into the sea. Subjecting a Jewish VP to the equivalent pressure of 12,000 metres of altitude, or hypothermic conditions that could stop their hearts, the tests took the subjects through a range of extreme states of life that would lead, for some, to death.

The documentation of Rascher's work, under the oversight of Himmler and Hitler's personal physician, Dr Brandt, was used as evidence during the Nuremberg trials to convict Brandt and several of Rascher's fellow scientists of war-crimes. (Rascher and his wife were executed by the SS two weeks before the Allies entered Dachau.) Investigated by Major Leo Alexander of the Medical Corps of the United States Army (Alexander 1945), the files emerged in Himmler's cave depository in Hallein, Germany. Now the study of research in medical ethics and the scientific dimensions of totalitarian regimes (Bogod 2004; Poszoz 2002), their evidence gives us explicit insight into the role and imagination of the aerial body.

Our concern here is for the reduction of the *Versuchpersonen* to a bare life. Viewed as non-living or already dead, the VPs were criminals or persons whose animation did not matter; they were 'asocial individuals and criminals who deserve only to die', as Himmler would come to justify their sacrifice.[3] Understood as little more than an organism, their use would be one of 'vital importance to the air force' and therefore national security,[4] avoiding 'a young German aviator' being 'allowed to risk his life'. The VPs tell us about the embodiment of aerial mobilities, the movement of the 'haughty' airmen who must be sustained in extreme conditions. But they also reveal bodies in the shadows, body-subjects quite removed from the aeroplane and the ways we tend to imagine their action.

1998, a cargo aircraft somewhere between countries

Almost 60 years later a series of air travels are experienced and written up by writer Barry Lopez (1999). Lopez had undergone 40 flights with air-cargo aboard various 747 freighters or passenger planes carrying freight.

Covering over 110,000 nautical miles, flying in and out of major cities and across the world's continents, Lopez visited Taipei, Rotterdam, Los Angeles, Lima, Calcutta and Chicago O'Hare airports, among many others. He travelled with a bewildering amount of cargo: from cattle and sheep to valuables such as precious stones and watches, to perishables such as flowers, food and newspapers.

Clearly this is not a *moment* particularly unique or worthy of drama, although Lopez's flights are full of little events (severe turbulence especially). His recordings capture the strange worlds of the cargo-hold, dominated by a stillness born out of animation: inattention from the overload of wind and noise; the stench of animal faeces or a fruit's unique odour. Lopez is both enthralled by the atmospheric little worlds of air-transport cargo and bemused by and critical of what they add up to. A museum director in Los Angeles, Lopez is told, found it cheaper for the museum's entire sandstone façade to be quarried in India before being air-freighted to Japan, where it was dressed, and finally 'flown to Los Angeles than to have it quarried, dressed and trucked in from Minnesota' (Lopez 1999: 84). Phone-books are shipped to China to take advantage of cheaper labour forces, who can key in the details at a far lower cost. Rayon blouses are cut in Hong Kong and flown to Beijing to be finished by hand before they are flown back.

Lopez's point is more subtle than a critique of the ridiculously generalized processes many of us call 'globalization'. Clearly it is a classic narrative of the annihilation of space by time (Harvey 1989; Kern 1983). It is a story of new economics, just-in-time delivery. Yet it is a far more human story than that. For the flights criss-crossing the globe, arriving in and out of cities, loading and unloading, are governed by desire. What planes fly, Lopez suggests, 'is what people imagine they want. Right now' (1999: 85). Flying over the rocket fire and streams of tracer ammunition in Afghanistan, the desire for guerrilla weaponry is nowhere more clear. Demand produced from a want, a need, a lust, jealousy, a most basic and thoughtless emotion, has produced a disturbingly uneven logic of aero-mobile commodity flows. The view 'that people everywhere want more or less the same things' is an illusory one; 'not all the world's cultures can be folded into [the] shape' of a European and Western consumer ethic (1999: 92). Stepping out of the homogenized airport in a place such as Calcutta or Harare, Lopez was witness to 'starkly different renderings of the valuable' (1999: 93).

Lopez's travels also brought him into the cockpit. With lengthy spans alone with the pilot to make conversation, Lopez asked him what he thought of the work of the artist James Turrell, who had built a crater in Arizona near Flagstaff in an attempt to construct what he thought of as the shape and volume of the air. Lopez's narrative cuts back to a recollection of a conversation with Turrell. 'For me,' Turrell explained, flying really dealt with these spaces delineated by air conditions, by visual penetration, by sky conditions; some were visual, some were only felt. These are the kind of spaces I wanted to work with. [...] People who have travelled to Roden Crater – heavy equipment operators as well as museum curators – say, yes, you do see that the sky has shape from the crater. (1999: 107)

After explaining the motivations behind Turrell's work to the pilot, the pilot duly 'turned around in his seat and said, "He's right. I know what he's talking about. The space you fly the plane through has shape."'

I asked if he thought time had boundary or dimension, and told him what I had felt at Cape Town, that time pooled in every part of the world as if in a basin. The dimension, the transparency, and the agitation were everywhere different. He nodded, as if together we were working out an equation. A while later he said, 'Being "on time" is being on fire.' (1999: 107)

* * *

So here we have, some 80 years apart, three rather different apprehensions of time, space, body and the air. For the geographers, the air allows one a greater appreciation of the ground from its high perspective; a greater albeit instrumental respect for geography, even as space is shrunk by the ability to traverse it; a new 'substance' of 'geographical knowledge' (Wright 1952: 330). From an enhanced position above activities and patterns on the ground, their perspective merges with many familiar narratives of flight as transformer of our senses of space and time, 'transcending geography, knitting together nations and peoples, releasing humankind from its biological limits' (Sherry 1987: 2). But geography was not transcended. Space, in its vertical and horizontal planes, is connected. The aeroplane depends upon the geography of the earth for it to survive. Reversing Gillian Fuller and Ross Harley's (2004) thesis, *life on the ground surely changes that in the air.*

As the geographers make a distinction between the body on the ground and the body in the air, Rascher's disturbing experiments make another sharp and literal cut. The choice is made between a body whose life is worth nothing – the *Versuchpersonen* –and the body whose life is worth more – the German pilots. The VPs' rights are suspended. They are exposed to harsh extremes of high altitude, freezing temperatures, the immersive conditions of aerial space. The issue here is the aeroplane's relationship with an inside and an outside, an insider and an *Other*. The *other* in this case is outside the realm of law, designated by sovereign and scientific power as 'asocial', 'criminal', and, therefore, whether they live or die is of no consequence. In fact if they survived the experiments they earned a pardon – they earned their right to life. The *Versuchpersonen* are in fact the aeroplane pilot's *alter ego*, the airman's *doppelgänger*. They are the *other*, both a by-product and an essential component of the pilot's operations.

Lopez's geographies of flying mark out other sorts of differentiations and productions of space. They are found unevenly in different places and at different times, and they are spaced by collective feelings. Together with Turrell and the pilot's accounts, the aeroplane reveals the shape and dimensions of the spaces it has produced: of unevenly created lines of desire along which commodities and aircraft travel; volumes of space where time is different; contours of expansive vistas, limits, borders and forms.

This book is precisely about the interdependencies identified by the geographers, the sharp and scary differentiations of bodies and subjects produced by the German scientists, and the spaces, shapes and volumes Lopez articulates – the solution within which aerial life finds its suspension. It is concerned with how space has been produced, transfigured and shaped through the technology of the aeroplane, and, as this has happened, how people have been changed too. If as Lefebvre famously suggests, 'to change life [...] we must first change space' (1991: 190), this book is about life changed and threatened by new productions of aerial space and mobility.

Overview

The spaces and shapes of the aeroplane are many and they are diverse. Airports and aircraft have become synonymous with our contemporary mobile world (Graham 1995; Hannam et al. 2006). More of us fly now than ever have before. International tourism is made possible by charter and scheduled aircraft flying between cities and continents. Western societies are made and constituted by air travel, allowing social relationships, networks and associations to be held and maintained, or, conversely, to be *dis*-abled, destroyed and ruined (Cwerner et al. 2009). War in the twentieth century was war waged by the aeroplane. From the air raids of the Blitz to September 11, to the newest unmanned drone aircraft deployed atop Gaza and Afghanistan, aeromobilities provide both promise and possibility as well as dread, terror, destruction and death (Grosscup 2006).

The spaces of the aeroplane, then, are very dangerous and they are certainly curtailed (Sweet 2004; Wilkinson and Jenkins 1999). It is in the spaces of air travel where societies are increasingly regulated as flight has become a dominant means to cross borders (Salter 2003, 2008). Flight is a space which is intensely segregated and hierarchical as well as highly monitored and controlled. The geographer's point is made nowhere more clear as the complex processes of securitization on the ground (in the airport) secure the way for a safe flight free from explosives, guns or hijackers.

On the other hand mobility by air is not nearly always so dark. The topside of its more concerning implications are the capabilities it brings to connect people together, to join friends, families and associates; to deliver aid and humanitarian assistance; to create exciting experiences and feelings of uplift, exhilaration and joy; and to create jobs, investment and value.

In many respects, aeromobilities are responsible for our current and modern condition and they are conveniently sought as the barometer of the day. They define and undo more traditional conceptions of citizen and territory through mobile post-national citizenship regimes or re-imaginings of nations. Airports' outlet-lined corridors express the relationship of consumer and commodity – brand names provide a welcome familiarity in a space that is often disorientating. The airport terminal is now even understood as the new model of the city. The *aerotropolis* or airport metropolis has become the future of urban existence, posited by John Kasarda (Charles et al. 2007; Kasarda 1991a) as the newest Kondratieff wave (Kasarda 1991b) of economic development, forming an air-cargo industrial complex essential to economies and especially city-states such as Singapore, Hong Kong, Dubai, and destination 'experience economies' (Lassen et al. 2009).

Clearly these issues demand investigation. Yet the study of aeromobilities has remained remarkably asocial and rather fragmented. There are entire journals devoted to aviation medicine. Aviation law even used to be a popular sub-discipline and is now a popular area of practice. Aviation and transport security is an enormous industry with well-funded research agendas. Air warfare has been subjected to its own scrutiny by those concerned with the effectiveness of payloads, or by international human rights organizations working on behalf of punished populations. Transportation research and transport geography examine air travel's relation with policy making, governance, economic transformations, alongside technological change. As we will explore in more detail in the following section, the way aerial mobility has been explored in the literature is as an instrumental device – it is a means to an end.

On the other hand, recent conceptual turns towards notions of 'mobility' and 'flow' have been productive in their elaboration of the mobile dimensions of existence, serving as some sort of corrective to the manner we have thought about movement before (Adey 2009; Cresswell 2001, 2006a; Urry 2007; Urry and Sheller 2006). More discussion and debate is also generating fruitful relationships between 'new mobilities' and a revitalized transport geography (Dival and Revill 2005; Knowles et al. 2007). While incredibly useful for the analysis of our aerial world, they tend to focus their attention to specific sites such as the airport terminal with convincing empirical detail (Fuller and Harley 2004; Pascoe 2001; Salter 2007a). Others use the airport as a metaphorical comparison to contemporary society (Augé 1995; Castells 1996; Chambers 1994; de Botton 2002; Serres 1995). The result of all of this is a useful, yet necessarily partial and often allegorical perspective on air-travel mobility.

This book tries to break the current mould. It aims to present theoretically informed research that explores what the development and transformation of air travel has meant for societies and the human subject. It argues that air travel is both constituted *by* and expressed *in* a set of geographies, infrastructures, relations and processes that connect both land and air.

Aerial Life

As a key component of contemporary mobile life, aeromobilities have shaped and defined the scope of our movements: the sorts of places we may go; the kinds of violence we may inflict; the scale, extent and manner of our surveillance (Adey 2004a, 2004b; Salter 2004). They alter the shape and sovereignty of the space above us, the way we see ourselves in relation to our country's neighbours, and the possibilities and capacities of our bodily movements. Gillian Fuller and Ross Harley (2004) go as far to suggest that airports are entirely new kinds of life. *Aviopolis*, they purport, is a 'mix of multiple forms of life' between the earth and the sky. As 'metastable' metaforms, airports 'mix multiple forms of life, matter and information into a series of new and constantly changing relations [...] it is impossible to separate airports from the ecology of its environs' (Fuller and Harley 2004: 104–5). The airport has an anatomy; it terraforms its ecologies – understood as a transversal exchange between nature, culture and technology – giving life to some and death to others. With its 'exoskeletons', circulatory systems of 'arteries' and 'capillaries', the airport is a 'commercial organism' (Harley 2009). In the same vein, International Relations scholar Mark Salter has noted how the operator of Frankfurt Airport City describes its passengers as 'the *homo aeroportis globalis*' – 'a new but by no means rare species' (2008: 11) – a distinctive kind of qualified and consumer-driven life, hybrid and adaptable, a species whose patterns of behaviour are difficult to predict.

Although we could take Frankfurt's claims with more of a grin than any acute interest, their ambition to extend and even replace the city articulates how quite new forms of life, exchange and political community have been produced by the aeroplane. I want to take these 'new forms of life' seriously in three related ways which I shall develop further and in more conceptual terms in the following section and throughout the volume.

1 Firstly, the increasingly central role of air-transport mobilities to our society has led to the genesis of what we could call an 'aereality' – a distinctive kind of mobile society, a 'life on the move' – which the aeroplane has worked to imagine, define and mould (Cresswell 2006a; see also Sheller and Urry 2000 on the car). At the same time, and as already discussed, the aeroplane threatens to destroy that life and other non-aerial forms of existence. Aerial life is thus a life on the move, on the edge, on the verge of disaster. We can, however, be more precise than this 'life on the move' given that the life-worlds of the aeroplane are unevenly distributed and they are not one but many. By this I mean not the separate societal incarnations of

the aeroplane or the aeroplane's uneven transformation of social structures and relations, but specific 'aerealities', in William L. Fox's (2009) terms – *ways* of 'life on the move' which the aeroplane has produced and threatens to undo (see, e.g., Gottdiener 2000). This book takes seriously the lives the aeroplane has altered.

2 It is the contradictions of these lives and indeed the contradiction of aerial life that this book is interested in. For if the city – the *polis* – transformed political life – 'from mere life to human life, and from human life to the good life' (Bull 2007: 13) – we must ask just what kind of life our aerial world has produced as it becomes increasingly the medium for the operation of violence, civil society, protest and political power. Life is both supported, shunted and made good by the aeroplane, and simultaneously dropped, punished and treated as less than human. Given Nikolas Rose's concern for our growing ability to 'control, manage, engineer, reshape, and modulate the vital capacities of human beings' (2006: 3), we must ask precisely how what we think of as life has become indelibly altered by our aereality. *Just what are the politics of aerial life itself?* The book explores how aerial life might be understood as a different kind of (non)human existence, altered in its capacities to think, feel and act.

3 Finally, this book is concerned with aerial lives as they are *lived*. Thus, it departs from the confines of artistic and visual representation such as cinema, painting or tourism that may characterize the study of aerial geographies or mobilities (Lofgren 1999; Paris 1995; Paschal and Dougherty 2003; Wohl 1994). In the plethora of ways that air travel has been examined, the representational portrayal of the aeroplane as a visual image and a subject of discourse dominates concerns. Focusing on the practised and performative dimensions of these aerial lives, the book examines the more-than-representational (Lorimer 2005) performance of aerial life to see how it is not only portrayed, but *done* and lived. Before exploring these conceptual concerns in any more detail, let me firstly illustrate the methodological shortcomings of attending solely to the representations of flight.

Research on the aeroplane is dominated by representation. For instance, Robert Wohl's (1994, 2005) wonderful epochal studies of aeroplane culture present the political spectacle of flight as an object of visual capital. Alan Cobham's arrival in England on the Thames. Lindbergh's landing at Paris. The Beatles' arrival at Liverpool. Hitler's stage-managed mobility across Germany by aircraft. Aeromobility conveys a personal and technological charisma to be visually consumed. Airline advertising (Van Riper 2004), posters, maps and calendars are discursively investigated for their communication of far-off places, the celebration of the nation, or the glamour and romance of the air. As well as celebrity and splendour, the shadow of the bomber (Bialer 1980) connotes fear and destruction, vulnerability and threat. Propaganda is high on the list of the aeroplane's functions. In his studies of the 'golden age' of aviation war films, which coincided with the 'golden age' of cinema, Michael Paris' (1993, 1995) shows how movies from Howard Hughes' *Hell's Angels* to Tony Scott's *Top Gun* display themes of heroism and force, daring as well as danger, the cult of the airman and their prowess of technological modernism and as a 'metaphor of national achievement' (1995: 205). The sight of the aeroplane is just as significant as the view from it. The aerial view, of course, was preceded by the technology of the skyscraper, producing already changed forms of 'aerial subjectivity' (Waldheim 1999), and constructing what Davide Deriu (2006) calls the viewing subject of the *planeur*. Skyscraper life meant living in heightened elevations where everything that was desired could be 'contained within' it or 'sighted from' the building (Douglas 1999: 202).

This book aims to move beyond the limits of discourse and textual representation in how we understand the aeroplane's worlds. It will investigate the aeroplane's manufacture of quite different ways of being. Inspired by Peter Sloterdijk's (2009) modification of Heidegger's 'being-in-the-world' to 'being-in-the-air', this book takes seriously aerial geographies as an 'unthought' phenomenal world. With the body and its 'vital capacities' (Rose 2006) as my starting point, the book complicates the predominance of visual interpretations – as merely scopic – towards a flesh-and-bones approach to visual culture – bodies that sense, touch, feel and intuit. It is therefore about aerial subjects and their relation with space, a realization 'that man is not only what he eats, but what he breathes and that in which he is immersed' (Sloterdijk 2009: 84).

It is not, however, about discarding representation or ignoring the power of the imagination. Investigating the production of aerial life means that one cannot dispense with the imaginative. Rather than pacify the visualization and imagination (Gregory 2004) of an aerially mobile subject, I suggest that these imaginations are bound up in the production of aerial subjectivities, in the production of aerial life itself.

Attuning to geographical imaginations and imaginative geographies, Derek Gregory's influential work on the construction of an apparatus of contemporary colonialism paves the way for an approach that considers how the imaginative geographies of colonialism are not merely an expression of processes and relations but actually do a lot of work. Inspired by Edward Said's *Orientalism* (1978), the geographical imagination forms more than merely an archive of cerebral thought produced through the sedimentation and accumulation of memories and histories; it actively structures future practices and performances. For Gregory, 'its categories, codes and conventions shape the practices of those who draw upon it'; in short, '*it produces the effects that it names*' (2004: 18). In this sense the imaginative geographies of *Orientalism* are closer to a kind of practice, a performance of space. It is thus a 'domain but also a "doing"' (Gregory 2004: 19).

Taking this cue, the book investigates aerial worlds as they exist as both real and imaginative domains, and as they are practised as both representational and more-than-representational performances. In this way the little things of bodily motion, technologies or technique produce or replicate minor or grand imaginative geographies, (re)producing imaginations of the body, the state, territory, the vertical, and so on.

But just as Said's imaginative geographies of the Orient presume an other, aerial lives produce and are premised upon outsides – constitutive others. In Agamben's (1998) study of Rascher's experiments it was the 'asocial', the 'criminal', a Jew accused of 'inter-racial' breeding, a Gypsy or a homosexual cast outside the rights of citizenship. Those were the Others to the security of the state and the safety of the aerial pilot. As well as the VP, the image of the Arab pervades the aerial imagination. The British strategy of colonial policing in Iraq first placed the Arab populations opposite the violence of the aircraft's bomb or machine gun (Lindquist 2000; Townshend 1986). Even T. E. Lawrence found the Arabs' mind-set difficult to understand. On occasion their response was 'too oriental a mood' for him 'to feel very clearly'. 'An Arab would rather offer up his wife than himself, to expiate a civil offence!' he explained to Basil Liddell Hart.[5] The 'interwar fantasy of the Arab' (Bourke 2005: 273) presented by Lawrence and British military officers is not entirely different, then, from the contemporary rendering of a 'sinister, bearded fundamentalist'. Such an image is embodied in the profiles and classifications of aviation security, where a 'physiology of seeing' (Pugilese 2006) casts an Oriental spectre of kinesiology, phenotypology and vestiture upon the mobile body. These *others* are identified, targeted (Gregory 2007), or produced through the lived techniques and practices of aerial life.

Aerial geographies

What I have developed is an argument as well as an approach. It is also more than an important or difficult methodological task; it is an empirical point. My contention is that the aerial lives I want to understand map a diverse yet coherent spatiality – an aerial geography – reminiscent of geographer Rachel Woodward's exposition of the multiple sites and spacings of 'military geographies' that are always 'there with us' (2004: 4) and James Sidaway's claim that the 'repercussions of war and death are folded into the textures of an everyday urban fabric' (2009: 1092). Take architect Eyal Weizman's (2002, 2007) critique of the 'territorial hologram' expressed through the Israel–Palestine conflict. Weizman (2002) argues that 'political acts of manipulation and multiplication transform a two-dimensional surface into a three-dimensional volume, thereby evading established models of spatial analysis'. For Fuller and Harley (2004) and Salter too, a space like an airport extends into both vertical and horizontal horizons, it moves 'into the atmosphere – integrated into transport and critical infrastructure, throughout government regulations and business relations, into the surrounding city and country, and across territorial frontiers' (Salter 2008: 9; see also Dodge and Kitchin 2004 and Graham 2004 on the virtual). The aerial lives I want to explore are three-dimensional too. They have volume, they take shapes and they frustrate ordinary forms of spatial analysis because they are littered and extended across an interlacing and interconnected architecture of both vertical and horizontal planes.

By taking the view from above and what Kenneth Hewitt (1983) called the 'view from below', the volume teases out something akin to what John Urry has labelled a system. Urry argues that air systems 'are central to the making of the new global dis/order' (2008: 25). The way I treat these systems or realities is slightly different. I want to explore the shapes and volumes these realities take and understand how they are formed and experienced in ways that threaten or secure the future of those systems. These are not that dissimilar to Lewis S. Thompson's description of a kind of aggregation, a medium of a 'boundless, borderless air mass', and an 'aggregate of men, aircraft, weapons, air bases, facilities and the industrial and technical resources that sustain flight' (1955: 58). Take bombing, a practice which has a history, as Gregory (2007) shows (see also Lindquist 2000), but it also has a geography with defining and competing contours and geometries. Understanding aerial geographies as a contradictory aggregation of spaces, shapes and geometries perhaps best characterizes my approach.

For example, consider Louise Amoore's (2009) discussion of the aligning of science, commerce, the military and the state, as she investigates the contemporary security apparatus or *dispositif* distributed across numerous terrains, materialities and systems of knowledge. What she calls an 'emerging geography of algorithmic war' embodies ever-closer relationships between 'science, expertise and decision', allowing the distinctions between them to 'become more malleable' (Amoore 2009: 54; see also Bousquet 2009). Evidence may be found in the smallest piece of computer code which expresses value-laden scientific models. It may be born in the movement of a body, portrayed in corporate sales jargon, or expressed in the instructions of policy. Three-dimensionally, Amoore's approach is to build up the aggregated shape of the contemporary security *dispositif* composed of numerous spaces, technologies, practices and knowledges.

Unlike the topological surfaces used to describe legal-judicial formations. expressed by International Relations theorist Didier Bigo (2006; see also Vaughan-Williams 2008) as a Möbius strip, the spaces I'm describing have corporeal volume and shape. Indeed, they are touchable, felt and immersive environs with form. On the other hand, they also create outsides. And like the Möbius strip, these outsides are relative to the positioning of the subject. In this manner they are more akin to Doreen Massey's (1993) sense of a 'power geometry'. For some, aerealities are exclusive and capable of exerting force and power. For others, cast outside or as *other*, they are exclusionary, pregnant with fear, and provocative of extreme forms of violence.

The book works through a series of time-spaces which are not chrono-logical or necessarily geographically delimited. Clearly the scope and limits of such a study could be absolutely enormous. Instead the book focuses on specific events, examples or case-studies in the history of the aeroplane's movement, particularly during its birth and examples from today which throw both into sharp relief, whilst preserving a genealogy of processes and trends. Although the majority of the original primary material I draw upon is European, American, archival and historical, the book uses diverse sec-ondary and supporting material from around the world, which actually often involves British and American forces or expertise, illustrating the imperialist character of aerial life (see also Aaltola 2005). The approach, however, is not intended to be distinct. It is not meant to speak only within the boundaries of the examples discussed, and they should be understood as exemplars or stand-ins for much wider processes. What's more, as the book tracks the different dimensions of aerial life, it sketches out the imagi-nation, augmentation, performances and experience of different subjects, be they passenger, pilot, scientist, planner, bombed civilian or artist, among others. These all surpass strict geographical and temporal definition. Clearly the volume cannot account for every body, and many will be missing. Instead I try to pick up on the subjects most transformed by the aeroplane's motion.

Aerial Life as both domain and doing, context and performance; before we move straight into the rest of the book, let us firstly explore the conceptual and disciplinary contexts and impulses that support this kind of analysis. How can we develop an approach more sensitive to the politics of aerial life?

Powering Up Aerial Geographies

We know how the study of our ability to move through the air has been slow to garner the theoretical benefits of 'new mobilities' and more sophisticated conceptual concerns. I do wonder, however, whether it is going too far to sug-gest that 'as objects of social-scientific mobility research [...] [airports] are almost entirely uncharted territory' (Kesselring 2009: 39), as does Saulo Cwerner when he suggests that 'aviation and air travel still receive scant atten-tion in the social sciences' (2009: 2). Numerous studies of airports and aircraft show them to be signifiers of our contemporary mobile world (Augé 1995; Castells 1996; Gottdiener 2000; Hannam et al. 2006; Pascoe 2001). Social relationships, networks and associations are held and maintained by what Claus Lassen (2006) and Paul Virilio (2005) first termed 'aeromobilities'.

Continuing work in transportation geography has served to trace out the complex linkages of airline flight networks and organizational patterns. Drawing out these vectors uncovers the elaborate relationships that connect certain places to others by air travel, pinpointing hierarchical structures of the most global and interconnected places (Graham 1995; Vowles 2006; Zook and Brunn 2006), supporting and facilitating international business relationships (Derudder et al. 2008a, 2008b; McNeil 2009). Other work on the social and cultural dimensions of air travel (Adey et al. 2007) has explored the relationship between air travel, identity and ways of belonging (Adey 2006; Corn 1983; Fritzsche 1992; Simmons and Caruana 2001). The airport terminal has become a focus for much of this thinking, drawing analysis of what it means to inhabit these spaces as sites of alienation, strange encounters and inequality. For others the airport can be seen as a place of home, relative stasis, dwelling (Adey 2008a; Crang 2002; Cresswell 2006a; Gottdiener 2000; Iyer 2000) and omnitopia (Wood 2003). Elsewhere the symbol of the aeroplane has influenced ways of imagining one's place in relation to the rest of the world (Pascoe 2003), acting as a builder of national identity and citizenship (Raguraman 1997; Wohl 2005). While the geographical relations of the aircraft cabin, or the aircraft cockpit, witness the performance of gender relations (Whitelegg 2005), Lucy Budd (2009) plots the spaces outside the aircraft window, examining the complex geographies of airspace management. Diverse and individually discussed spaces are evident, yet the shape, scale and quality of these spatialities have received little attention and neither have their contradictions.

An earlier paper by political scientists Stefan Possony and Leslie Rosenzweig captures some of these shortcomings. Attempting to set out an investigation of the 'geography of the air' as a new research focus, the authors limited their scoping of this geography to the 'physical differences of the air in various locations and altitudes' (Possony and Rosenzweig 1955: 1). They wanted to know what 'that invisible sea in which we live, the air' (1955: 1), meant for international relations, strategy and foreign policy. The authors made some interesting observations and should be applauded for stimulating an awareness of the importance of air-geography or 'aerogeography'. Although their conception of this geography was rather unsophisticated and problematic in relation to today's studies, their approach was to make the kinds of connections this volume strives to make. Setting out how the physical characteristics of airspace constrained and enabled aerial activity, from the simple implications of air currents, altitudes and temperature, their environmental determinism turned the layers of the atmosphere into variables to be entered into the equation of complex geographical and social relations. My bugbear is that today's studies fail to escape the one-dimensionality Possony and Rosenweig overcame. They miss the depth of the sky, the infrastructures that support it and the social relations and experiences they make possible.

Cwerner, Kesselring and Urry's recent collection *Aeromobilities* (2009) points out that enhanced collaborative research can enable aeromobilities to be understood in this kind of way, 'by bringing together geographers, sociologists political scientists, semioticians and even artists, among other professionals in the humanities and social sciences' (Cwerner 2009: 18). As the editors suggest collectively, 'Aeromobilities calls for a collective effort to expand our knowledge of airspaces and the social relations they enhance and make possible' (Cwerner et al. 2009: x). These new and integrated research agendas enable more socially nuanced and interdisciplinary positions on a subject like air travel. Indeed, it is through diverse conceptual perspectives that the three-dimensional shapes and scales of aeromobility may be revealed, necessitating, I contend, far more conceptual sophistication if they are to deal with what we can call some of the really 'critical' questions that aeromobilites research demands. As Caren Kaplan has recently claimed, the topic of violence remains incredibly absent from conceptions of mobility, thus, 'considerations of the entangled histories of war and mobility serve as an important corrective' (2006a: 395; see also Virilio 2005). What are the 'quality of life' issues at stake here, of fear, feelings of vulnerability, the rough touch of power? This compels a more nuanced consideration of the body as it is experienced and performed, and more careful understandings of power and control.

Circulations, species and spaces

'[O]ur relationship with aeromobilities is deeply *political*', argue Cwerner, Kesselring and Urry. 'And, without knowing what lies above us, we have very little scope for bringing it under democratic control' (2009: x, emphasis added). Clearly, the complex geographies of extraordinary rendition, aerial bombardment or airport security require just as complex theorizations of power in order to understand not only their composition, but also their implications. Let us take the monitoring by plane-spotters of secret CIA flights, which has given us a new window into the clandestine world of movements in the sky (Grey 2006; Paglen and Thompson 2006).

The CIA-leased private jets were used in a system intended – and many would argue misguidedly and illegally – to secure everyday life, a life which had been threatened by the hijack of several commercial aircraft on September 11, 2001. It is a question of the circulations that aerial life produces and becomes threatened by. We well know how the aeroplane has allowed us to become ever more mobile, shunting the sorts of people and commodities described in Lopez's narration. We have already discussed the manner in which this has engineered quite new forms of spaces and bodies into what Urry would describe as a system of circulation. And yet, these circulations have become both the target and medium – the infrastructural vulnerabilities – of aerial life. The problem is therefore centrally a matter of security. For if the aeroplane, its mobilities and infrastructures are harnessed as a means of security, how are these techniques further productive and destructive of different kinds of aerial life itself (see Lobo-Guerrero 2008)?

Michel Foucault's (2003, 2007) writings on power and biopolitical security deeply influences my approach. A plethora of research that orbits around
Foucault follows the way mobility has become a problematic of security,
and, thus, conceptualizes biopolitical administrative techniques as practices
which take mobility as an inextricable component of what it means to live.
The biopolitical apparatus is therefore about the facilitation of circulation,
allowing things to be in motion, 'constantly moving about, continuously
going from one point to another' (Foucault 2003: 65), all with an inherent
danger that must be quashed. Circulation of all kinds – biological, monetary, trade – appears a contingent variable in the constitution of a free, good
and constitutively 'aerial' life.

A shift from a sovereign geopolitical engagement with territory to bodies
and populations marks Foucault's identification of biopolitical versus sovereign power. For many writers, Foucault's (2003, 2007) texts on the biopolitical help us interpret how contemporary forms of power enact a managerial,
and not necessarily disciplinary, enrolment of a molar – a group of people
addressed as a population. Bodies, and knowledge of bodies, are abstracted
into data which may be sifted, tabulated and searched with the aim of admin-
istering that population. Indeed, the continual calibration of circulation is
organized in order to protect that circulation from strangling life.

The key for the scholarship of Mick Dillon and Luis Lobo-Guerrero
(2008, 2009) is the precise object of these techniques. They ask what sorts
of technologies and techniques address the referent object of biopolitical
security – that is, life – or the 'species being' as defined by Foucault. And, if
what counts as life alters, what does this mean for technologies of power. It
follows, then, that if the aeroplane has led to the imagination and production of quite new kinds of life – or forms of life mediated by the aeroplane's
mobility that continually pose a threat to that life – *what sort of technologies
and techniques have led to its production and security?*

The response the book takes is with less fidelity to Foucault, but a more
considered take on biopolitics, sovereign and disciplinary power. Firstly, and
this is certainly not unique to my approach, in the context of the aerealities
I pursue, biopower is not necessarily divorced from the exercise of other
modes of force. In other words the modes of power discussed are typically
multiple. Gregory's (2008) investigations of the 'war on terror' exact a similar blurring of biopolitics with the 'protracted struggle over the right to claim,
define and exercise sovereign power' as demonstrated in the struggle over
insurgents in Baghdad. Similarly Colleen Bell's analysis of contemporary
surveillance and security techniques is suggestive of forms of subjection '"at
a distance" and compelling modes of self-governance on the one hand'
whilst, paradoxically, it evokes 'the presence of sovereign authority in modern
affairs of state' (2006: 151), on the other. The evolution of power from sovereign to biopolitical, from the focus on the individual body to the collective –
the discipline of the body and its capabilities to the collective issues of 'births

and mortality, the level of health, life expectancy and longevity' and the environmental variables that controlled them (Bull 2007: 8) – is not necessarily singular or linear in the formation of aerial geographies. Sovereign, disciplinary and biopolitical modes of power overlap one another.

Furthermore, as Achille Mbmeme asks, is 'the notion of biopower sufficient to account for the contemporary ways in which the political, under the guise of war, of resistance, or of the fight against terror, makes the murder of the enemy its primary and absolute objective' (2003: 12)? Moreover, 'What place is given to life, death, and the human body (in particular the wounded or slain body)? How are they inscribed in the order of power?' (2003: 12). Thus, what other theories of power can account for the decision to obliterate, wound and depress the enemy as a political objective? In the following I turn to conceptions of embodiment.

Finally, I dismiss the tendency to see biopolitics as aspatial. Foucault's three apparatuses of power – sovereign, discipline and security – are each applied to specific degrees of spatial prowess (2003). Sovereign power is applied over a field of territory, discipline to the body, and security to a population. Whilst the tendency of writers such as Agamben and Dillon has been to downplay space in favour of topology, others such as Gregory (2008) dispense with the container-like presuppositions of geography in favour of performances of space which are fluid, dispersed and temporary. What's more, if these different modes of power overlap one another, then questions of the population may be accessed through a specific spatial regime, migration movements may threaten the integrity of territory, the policing of mobilities by biometric technologies may be performed through particular disciplinary apparatuses and technologies of position. As discussed earlier, the *others* that aerial geographies produce are constructed through both abstract and material spaces that are positioned to be *outside*: both outside strict definitions of power and outside regimes of legal protection and cultural norms. These geographies are lived and embodied and they raise further questions for the technologies and techniques that secure them.

Affect, knowledge and material geographies

Attuning to the multiple and political spaces of aeromobility requires an approach far more sensitive to the embodied geographies of mobility, or an attendance to life on the move as it is lived. The mobility literature has been particularly slow in accessing phenomenologically sensitive concerns for the body-in-motion or -action (Harrison 2000; Seamon 1980). Bodily perceptions, feelings and sensations unearth a terrain of mobility-in-action that is difficult to portray or represent. In the context of automobility, for Peter Merriman, driving along the freeway is approached 'as a kinetic, non-representational, performative engagement' (2007: 15; see also Laurier 2004).

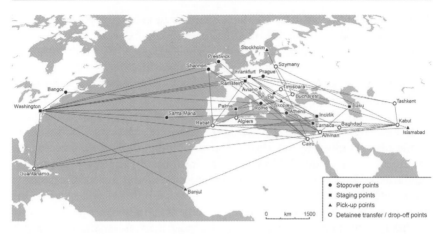

Figure 1.1   Networks of extraordinary rendition. (*Source*: J. Sidaway, 'The dissemination of banal geopolitics: webs of extremism and insecurity', *Antipode* 40 (1), 2008: 2–8. Reproduced with permission of Wiley-Blackwell.)

Keeping Sloterdijk's (2009) 'being-in-the-air' in mind, the approach I opt for takes the aerial body's 'being-in-the-world' in the manner that Tim Ingold (2005) has re-theorized the weather. Like Sloterdijk, Ingold focuses on the divisional artifice between the earth and the sky: bodies rendered as 'stranded on a closed surface' instead of being 'immersed in the fluxes of the medium, in the incessant movements of wind and weather' (2005: 534). Where there is no distinct surface separating life on the ground from the air, I take the notion that aerial life is submerged in the medium of the aerial, 'borne on these fluxes which, felt rather than touched, permeate the inhabitant's entire being' (Ingold 2005: 534). Indeed, the contradictions that continually threaten aerial life are bound up in this 'unthought' domain (just as Sloterdijk identifies the aerial environment as the primary object of terror). Attuned to the vocabularies of emotion and affect, the book runs in the direction of an increasing body of research which investigates the vitalist demand for animation (Thrift and Dewsbury 2000). That is, it develops the inherent movement and complexity of the world or milieu within which life is made and the 'vital capacities' that make up just what it is to be a living thing (Rose 2006).

Once more, take the study of extraordinary rendition. Also known as expedited removal and previously approved under the Clinton Administration, George W. Bush's war on terror used rendition in order to deal with people suspected of terrorist activities. Trevor Paglen and A. C. Thompson (2006) reveal networks of extraordinary associations – the mobilities of rendition – which deposit crucial traces. Usefully projected by James Sidaway (2008) into a 'web of extremism' (Figure 1.1), his map displays flight paths taken by Gulfstream jets hired by the CIA from innocuous civilian contractors based at just as innocent-looking small-town airfields.

Just this kind of representation of movement as lines on a map is chosen by geographer Tim Cresswell (2001, 2006a) to be the leitmotif of a metaphysics of movement that has dominated the study and understanding of mobility: abstract, linear and divorced of meaning – 'a movement from a to b' (Kaplan 1996; Urry 2000). Although the rendition map is not entirely devoid of meaning – it is actually quite startling and shocking if one considers its implications – the detainees forced to endure these mobilities tell a different story entirely. They tell of life on the move that is meaningful and affective. Put in chains and foot shackles, Maher Arar was taken to a grey shed on the grounds of New York's JFK Airport without being informed of his plight. In a room with metal benches he was deprived of sleep and found himself 'very very, scared and disorientated'.[6] Humiliated and taken to a jail for a few days, Arar was then driven to an airport in New Jersey and boarded a Gulfstream Jet, where he was chained to leather seats whilst his guards watched in-flight videos.

In contrast to the luxurious mobilities of his captors, Arar's movement was filled with anxiety and anticipation; it was a humiliation of (Mutimer 2007) strip searches, swearing and interrogation – all without proper legal representation. It was painful too. The conditions of his movement were restricted, shackled. The end of his passage to Oregon, Rome, Amman, Jordan and finally Syria resulted in more visceral physical and psychological torture – an intensification of his embodied suffering of mobility – that constituted what Matt Sparke (2006) defines as a form of neoliberal 'Gulfstream sub-citizenship'. Even the representations of passage that helped Arar to identify and confirm his movements were performances of emotion, affect and kinaesthetic sensibilities of (im)mobility. They not only helped investigators track Arar's passage, but their diagramming also reminded him of the experience. He found it 'painful' to remember as he relived watching the video screen displaying the aircraft's vectors and position whilst in transit. Arar could see and anticipate his travel as it happened, warning him of his imminent arrival.

The experiential affective and sensuous mobilities described here are what I want to get at. The aims of the Futurists who were so enamoured with flight were precisely the same: to understand and describe 'not only what they felt but how it also affected them' (Bohn 2006: 209).[7] Such a focus is receiving accelerated attention in the literature, which is experiencing some kind of 'affective turn' (Ahmed 2004; Anderson 2004; Anderson and Harrison 2006; Bissell 2008; Brennan 2003; McCormack 2003; Massumi 2002; Sedgwick and Frank 2002). 'Affects' or affective life is becoming increasingly the object of new forms of security and governance, as Mick Dillon and Julian Reid notice in the US government's war on terror: 'the movement of human beings, each and every possible human disposition and expression, of each and every human individual, is becoming the object of strategies for the liberal way of war' (2009: 140). Rose's 'vital capacities' are thus objectified and harnessed as hinge-points to the security of liberal regimes.

The aerial lives this book unpacks are therefore produced by techniques and technologies of which (aerial) life – in its affective and emotional dimensions – is their referent object. In the case of Maher Arar (and others), it was precisely this level of life at which his captors and torturers wanted to impose their presence. As geographer Ben Anderson writes, drawing attention to our 'capacities to affect and be affected' is to reveal the relations that 'enact the *life* of everyday life' (2006: 734). Antonio Damasio's (2003) continuum of affects is particularly useful, as Engin Isin (2007) has explored, demonstrating how power can be directed towards a host of affects that range from the most primordial capacities and primitive instincts and drives towards semi-conscious feelings and more qualified emotions at the other end of the spectrum. Far from an 'inaccessible substrate' (Foucault 1977), these bodily affects are addressed in several ways, 'holding together both disciplinary and biopolitical regulatory mechanisms, along with command in its sudden flashes of sovereign power' (Ticineto Clough et al. 2007: 73).

Establishing this register as an object of power or securitization requires *knowing* and practices of establishing knowledge, allowing affect to be 'passed into knowledge's field of control and power's sphere of intervention'; as Mick Dillon puts it, 'there is a mutually disclosive belonging together of power and knowledge' (2004: 54). Aerial spaces are overshadowed by knowledge practices which seek to understand and govern aerial bodies at social, meaningful and cognitive levels and at much more biological and affective registers. Knowledges of affect are thus incredibly important in this book's understanding of an apparatus of aerial technologies and techniques that seek to govern behaviours and futures based on apriori experiential knowledge and expertise. Attending to this register, therefore, marks a wider turn in the epistemological question of just what it is to be secured, or, moreover, just what it is that becomes the object of security and surveillance strategies (see also Dillon 2004). In the construction of these new epistemic objects (Rheinberger 1997), the referent object of the security of aerial life becomes precisely the subject that needs securing as well as the object that provides the threat to security (Dillon and Lobo-Guerrero 2008).

Addressing the enrolment of rather different kinds of knowledge, particularly from the social sciences (Agamben 1998: 159; Foucault 2007), the book will deal with different kinds of knowledge and the different degrees of efficacy and imaginations of the life they seek to explain and control. Aerial life will be understood as ecological, cybernetic, genetic, physiological, energetic, affective, sensuous, quantitative and more. Indeed, for Dillon and Reid (2009), it is increasingly recognized as *technological*. By tracing the nodes, networks and spacings of affective and expert knowledge, we can begin to lay out the shape and volumes of our aerial domain.

The Organization of the Book

The book is broadly divided into three sections which divulge different elements, tones, contours and spaces of aerial relations. Each chapter may be read independently of one another, whilst they follow threads and connections throughout the book. Chapters will build up specific associations and conceptualizations of our aerial lives which will inform the relations and processes discussed in the other chapters.

Part I, 'Becoming Aerial', is precisely about the bodies, spaces and performances produced through the aeroplane's relation with the nation, state and territoriality. Chapter Two starts with the body of the youth, the aerial subject being born as ready and militarized. It asks, precisely, how did bodies become trained to be aerial? In the context of the build-up to the Second World War, the chapter follows an analysis of youth movements formed with the aim of stimulating aerial enthusiasm among young people so that they would one day be able to perform the function of defending the nation, whilst inculcated with the values of good citizenship. The chapter sets the basis for our comprehension of aerial mobilities as embodied performances, doing citizenship, acting out and developing their nationhood through mobile actions.

Chapter Three identifies the aeroplane's involvement as a nation builder by examining the political spaces produced through its performance of (inter)national power. From the airshow to the airport terminal and the airspaces above them, the chapter investigates several choreographies of nationhood and sovereignty. As the aeroplane reinforces the container-like definitions of citizenship and territory, we examine their fundamental contradictions. Telling the narrative of remote borders and post-national citizenship regimes alongside no-fly zones and humanitarian crises, we see how the aeroplane's mobilities threaten the vital integrity of 'territory', 'border' and 'citizen'.

Part II, 'Governing Aerial Life', is self-evidently and more tightly focused upon how aerial life is governed and governs. Chapter Four is concerned with how the way we see *aerially* has worked to produce aerial subjects and spaces mediated under a field of control. Dwelling upon two kinds of seeing which are inconceivably comparable – aerial photogrammetry and biometric security – the chapter follows their processes and practices of *visioning*, drawing examples from the birth of aerial photography in colonial expansion to the control of mobile populations in today's international security regimes. Moving from abstract perspectives on the population to higher resolutions of the body, we see the creation of aerial subjects capable of being governed, of populations suddenly available to the scrutiny of colonial administration, or cleared for international mobility.

Chapter Five drops down away from the synoptic view of the population to aerial subjects visualized by their capacities to do action, opening up the bonnet – as it were – of the aerial body. Assessing the testing, training and governing of the aerial subject, the chapter investigates examples ranging from the experimental spaces of British and German scientists during the Second World War, to the contemporary security checkpoint. Whilst the practices of Chapter 4 were shown to be considerably rational and calculative, the chapter argues that the governance of the aerial body is increasingly attuned to the post-rational intuitions, urges and instincts of the body. It is at these registers that the aerial body can be sorted, distinguished and excluded, as well as improved, trained and produced.

Part III, 'Aerial Aggression', the final section of the book, concentrates on violence: the aeroplane's ability to inflict it and the population's capacities to withstand it. Chapter Six focuses on the environments imagined and created by the aeroplane's aggression, beginning with the formation of bombing policy during the British colonial policing campaign. In the study of science and the waging of aerial war, the chapter explores how the target of the aeroplane has been displaced. Unpicking the relationship assumed between the body and its environmental milieu – from colonial counterinsurgency in Malaya to today's operation of 'Shock and Awe' – we examine how the target of aerial war has found its location in the environmental conditions upon which life depends to survive.

Chapter Seven picks up where Chapter Six left off, beginning with the wailing sirens administered to historical and contemporary populations on the precipice of aerial attack. A mirror of Chapter Six, the chapter explores how aerial violence should be understood as *symmetrical*. Utilizing the same techniques of aggression as protection, aerial populations are defended by environments conceived with the purpose of sheltering the body at the level of its biologic and affective capabilities. The chapter dwells on a continuum of aerial protection by investigating – in detail – the abilities of civilians to prepare and anticipate bombing during the Blitz of the Second World War, which is paralleled with today's preparations for the *event* of a terrorist attack.

Part One

Becoming Aerial

Chapter Two

Birth of the Aerial Body

[T]he lessons here are biting deep, and we shall never be the old selves again. Does it not rather frighten the R.A.F. to remake so many men after its desire? Bodily we are being built to drill-book pattern: spiritually we are being moulded nearly as fast.

(Lawrence 1936: 28)

[A]ffects [...] are basically ways of connecting. They are our angle of participation in processes larger than ourselves. With intensified affect comes a stronger sense of embeddedness in a larger field of life – a heightened sense of belonging, with other people and to other places.

(Massumi 2003)

Introduction

Consider a speech delivered by Captain F. E. Guest during his half-yearly inspection of the Royal Air Force (RAF) officer training station at Cranwell, Lincolnshire, during August 1922. Guest stated,

> I confidently expect great results to follow from the infusion of what I may call the true Air Force blood, as our units are gradually permeated by officers passing out from this College who have been taught from the beginning to think in terms of air and not of sea or land. [...] You who are here to-day will be pioneers of the science of air warfare – a new race of men who will have leant to think and move freely in three dimensions. For many centuries past soldiers and sailors have been confined to the narrow limits of the horizontal plane; you airmen are mastering the vertical, and will have a freedom of movement of which your predecessors did not dream.[1]

Albeit extraordinarily romanticized, Guest builds an image of bodies being made to be aerial. The pioneers were to step beyond the horizontal plane and to become *masters* of the air. Guest speaks of officers who are under construction – a new race, or organism in fact, who are being taught to think, act and move in a world of different dimensions to their predecessors. A new breed of aerial or air-men for a new age of freedom and a new science of air warfare. Infused in the Air Force blood, the spirit of the officers should permeate into every pore of the new recruits. Pulled kicking and screaming into flight, the aerial subject will achieve mastery of the new worlds flight creates (see also Corn 1983).

This chapter explores the establishment of aerial bodies – produced and manufactured bodies that had to be designed, preened, screened and developed into aerial subjects whose destiny it would be to secure and defend the nation. It was already too late for those whose bodies had developed: their habits were routine and beliefs too solid. Organizations interested in turning out these subjects realized that a body already in development could be tailored and shaped to fit the mould of an aerial or *air-minded* subject-citizen.

This chapter avoids lingering solely on what flight meant to young people at this time. *Venture Adventure* went the motto of the Air Training Corps (ATC), a phrase that was meant to be said aloud. Like the spoken word, the aerial body was an utterance.[2] An aerial life was to take shape through various processes of training and character building; it was not an idea to be activated by the flick of a switch. Bodies could be developed by organizations such as the RAF or the Luftwaffe, the ATC, the Air Scouts (a branch of the Scout movement), and in national youth movements. If the public were to be aerial, then they had to *become* it; they had to be made aerial through a host of different ways of *doing*.

The performances of public displays and airshow demonstrations were intended to inspire and encourage people in the promise of the aeroplane. In the same way, parades and public drills performed by the bodies of aerial organizations were intended to demonstrate and communicate air-mindedness through their disciplined mobility and choreography. Air-mindedness could be instilled in the wider public. But it is not symbolic or discursive communication that I want to focus on in this chapter. Nor were these symbolic displays the only way people were made interested in aviation. Rather, I want to investigate the sorts of mobile practices and actions people were encouraged to perform. Although entrained bodily movements were part of the parades and demonstrations, the public flag-waving spectacle can be seen as a useful outcome but not their underlying intention (Lorimer 2005).

The examples I present emphasize action; they privilege mobility over immobility. It was believed that the bodily actions the youth organizations endorsed and displayed would contribute to the formation of desirable moral, intellectual and physical qualities or capabilities within the youths who performed them. As youths strove to become aerial, they would be ordained with these qualities. To borrow from Michael Rosenthal (1986), the organizations became the 'character factories' of the air-minded (see also MacDonald 1993; Mechling 2001; Springhall 1977).

Focusing on 'character', the chapter follows the inculcation of the air – an *aerial life* born through sets of associated practices of the mobile body which had their own benefits in the training of character and, importantly, the 'capacities' desirable for their militaristic use. The idea of character was made sense of as much more than some sort of mental or psychological quality (Gagen 2004, 2006). The chapter traces how these practices invested qualities at an embodied, material-affective domain of instincts, drives, feelings and emotions which were inextricably linked to desirable values and ideologies, especially of citizenship. Believed to be composed in the physical training of the mobile body, such practices drew upon a context of a physical kinaesthetic education, which would construct patriotic and model soldier-citizens – rebelling against Victorian classroom-based oratory learning.[3] In combination, these techniques could provide an effective concoction that would train, educate and mould susceptible bodies in patriotism, citizenship, action and all things of the air.

As with David Matless' (1998) discussion of the almost flow-sheet-like process of the pre-Second World War Scout system, which would turn ordinary boys into useful citizens of the Empire, a processual sequencing was employed to make youths aerial. Take a similar flow-sheet diagram that was drawn up in the planning of the branch of the Air Scouts.[4] The chart was organized by outlining the perceived benefits to the nation and the individual scout's character. The first main aim was to 'honour the RAF'. This would be achieved by instilling a sense of 'honour and reverence' in the youth, and a 'showing of their spirit' in practice. Secondly, the Air Scouts would provide an 'air-minded nation'. This would occur by educating scouts' minds in the knowledge of 'men and machines', while in practice they would make efficient agglomerations of muscles and nerves through training and drilling. Thirdly, it was the aim of the Air Scouts to 'produce useful citizens' for the nation by instilling a 'willingness to serve' in mind, and in practice a preparedness echoing the Scouts' motto 'Be Prepared'. Finally, the Air Scouts could fulfil the needs of the nation by encouraging their scouts to know more in mind, and importantly being able to 'do more' in practice. This character-forming process could have considerable benefits for the boy as well as the nation. According to the chart, it would 'teach him to be resourceful', and it would 'serve and direct his natural urge to hero worship'.[5] This could be achieved by giving the scout knowledge of famous flyers and pilots, and 'in practice' through 'imitation'. In went the youth. Add various activities and practices to their training, and out came fit, willing, responsible and readied citizens of the Air Age.

Firstly, the chapter follows how the aesthetic of the aerial body was disassembled. The heroic and handsome aviation ace had permeated popular culture, particularly abroad. All but captured in the face, this model was eschewed in favour of a culture that looked both behind and below what sat on top of the shoulders, valorizing embodied action rather than an idealized body image. Secondly, the chapter follows how the characteristics of an aerial subject were developed by the exposure of the body to the world of aviation. From the ATC hut and the Lincolnshire countryside it was supposed that the exposure of the mobile body would develop an 'air sense'. Thirdly, the chapter examines how the development of the bodies-in-formation could help instil air-mindedness through exercises and practices which were aimed at conditioning the 'muscular consciousness' of the body. As shown in the final section, these included literal performances of aviation-related scenarios and simulations through carefully choreographed mobilities.

Beginnings

Efforts to encourage people to become aerial were initiated in Britain by the Air League of the British Empire. In France and elsewhere, similar pressure groups and political parties drove home the importance of the air through several common elements. French historian Sean Kennedy describes these as '[m]ass mobilization, national regeneration, unity, discipline and anti-parliamentarism' (2000: 384) – aspects all successfully integrated by the Croix de Feu, the largest of France's interwar nationalist leagues, into a vision of air-mindedness.

The Air League had attempted to encourage the involvement of the Scout Association in all things aviation. They did this by inviting them to flying meets and races, and encouraging the Scout Association's publications to feature articles on famous airmen and their exploits. The Air League also set up a sister organization in the form of the Women's Aerial League, and the first real organization aimed at fostering youth to be air-minded can be found in the Young Aerial League. Comparative organizations were created in Germany through the gliding prowess of the Hitler Youth, as Peter Fritzsche (1993) shows, and in the United States, most notably the Air Youth of America, financed by Laurence S. and Winthrop Rockefeller from Rockefeller Plaza in New York. Young people were the susceptible clay from which a new kind of aerial being could be moulded, and the seedbed for a new form of aerial and political community. Scott Palmer's (2006) deconstruction of a Soviet postcard presses this home even further. The postcard, produced by Osoaviakhim in the 1930s, portrays a peculiar composition of Stalin holding a child in his arms, a woman breastfeeding an infant, and – in-between – three aircraft swooping from the sky. With the slogan 'We respond to Stalin's concern for children through a ceaseless struggle for the life and health of every child', the postcard sings heavy overtones of 'father-Stalin, mother-land and pilot-son'. The youth was where aviation should begin, breathing life into the national community.

In Britain, the progress of the Scouts in the air was led by the efforts of their founder Robert Baden-Powell's brother, Major Baden Baden-Powell (Jeal 1989). As a member of the Royal Aeronautical Society, Baden

Baden-Powell was convinced of the importance of the aeroplane for warfare and imperial security.[6] The Air League's attempts to instil an 'air sense' among the country's young produced propaganda material such as their *Facts about Flying and the Civil Uses of Aviation* (Air League 1923). The book featured detailed information about the potential for the technology of the aeroplane to ensure the security of the British Empire. Some 4,000 copies of the book were bought by the Scout Association alone.[7] Through the wrapper of education, the Air League's message found its way into schools, who introduced the text into their libraries, although the League did face some resistance from the London County Council, who neglected to adopt the book. For the Air League this rang of a 'distorted mind' obsessed with pacifism.[8] They even went on to suggest that by neglecting to purchase the book and educate the children on flight, the Council would be failing to allow the child to mature and develop as he or she could. The League proposed that 'the mind of the child that is taught in this manner will not be properly developed, either as regards the historical facts in the past, or, what is far more important, with regard to the deductions that ought to be drawn therefrom'.[9]

While the Air League found it difficult to penetrate the school syllabus, for their counterparts in other countries this was far from the case. Germany was able to introduce aerial matters into many aspects of the school curriculum. The British Air League soon sought other means. The League was present at exhibitions such as the 'Boys Own', held in London in 1926, and set an essay competition asking youths to discuss the 'Importance of Air Communications for the Development of the British Empire'.[10] The winning essay was published in the *Air League Bulletin*, and contained an imagined journey through the Empire by aeroplane.[11]

However, capturing people at this early stage was one strategy, but how to follow through this development into an aerial adulthood was quite another (see Chapter Five). By the mid-1930s the League were operating schemes on a much more practical basis as it became clear that a radical and direct involvement was needed to awaken the air sense of Britain's youth. The Young Pilots Fund was developed to provide bursaries to help young people to fly. Nevertheless, in comparison to other countries the progress of the Air League was slow. The Nazi disciplination of air-mindedness had enabled 'virtuous lessons about self-reliance and patriotism' to be taught through the special emphasis on gliding. From textbooks to the schoolroom a new 'intellectual scenery' emerged, of 'model airplanes hanging from the ceiling; air war murals on the walls; dozens of airplane books on library shelves' (Fritzsche 1992: 203). Moreover, by 1939, the US Civilian Pilot Training Program would take off (Pisano 1993).

The Air League struggled to impassion those who would listen. They argued that the country's security depended on making the young air-minded.[12] Aiming to put a pilot into every village and accompanied by the slogan 'HELP THE AIR LEAGUE TO HELP OUR YOUTH INTO THE AIR', the Air League tried to raise money for the cause. The promise of the pilots' fund was promoted alongside the potential for the wider deployment of air-mindedness across the country as a whole. In a letter to Lord Brabazon of Tara, the Air League requested the airman's support for the scheme on the basis that 'each of these young pilots will form a focus of enthusiasm among his relatives and friends so that we shall get that support for aviation and recognition of its possibilities which is as essential for our prosperity as for our safety'.[13] Getting one youngster to fly, it was believed, would snowball into wider family networks stimulating even more to support the 'air' cause.

As war loomed in 1938, the efforts of the Air League rapidly accelerated. The idea of the Air Defence Cadet Corps (ADCC) was conceived as a youth movement with the mandate of eventually recruiting boys aged 16–18 into the RAF as pilots or aircrew. '[S]martly uniformed and disciplined' and instructed in 'air matters',[14] the Air League aimed 'for numbers of youngsters definitely trained to take an active part should war break'.[15] As the ADCC tried to set up units attached to schools, they came across a certain amount of resistance from those who saw the threat of militarizing youth groups, especially the League's old adversary, the London County Council, whose task, they argued, 'was to educate the children, and not to turn them into soldiers'.[16]

By 1941 the ADCC had recruited some 20,000 cadets. The government, in realizing the considerable need and potential for air-minded youngsters, took over the scheme with the intention to expand it on a vast scale. The ADCC were renamed the Air Training Corps in 1941. Fed by the Air Scouts, the Corps began recruiting cadets with the intention of boosting numbers to 120,000 within the year.

'Handsome Is as Handsome Does': Disassembling the Aerial Body

Idealized bodies permeated the early characterizations of aircraft pilots in fiction and popular culture, reflecting wider ideological constructions of a heroic and patriotic physicality. These images had to be undone.

For Valentine Cunningham, the 1930s were characterized by the universal motifs of 'airmen, mountaineers, mountains, eagles, leaders, aerialism, and so on' (1988: 155). The aviators' heroism was emphatically linked to an 'era of the body'. The first pilots were awarded chivalric and heroic status through their actions and deeds. They were ordained with the status of a reckless adventurer fulfilling the dreams boys read about in magazines such as the *Modern Boy* and *Popular Flying*, blurring myth with aesthetic bodily form. The spectacle of flight conjured Hollywoodized fictions of 'helmeted knights wrapped in white scarves and joisting in the air with blazing machine-guns' (Wohl 1994: 204). The image of the ace was a tall, muscular male who for Wohl 'combined the daring of the acrobat, the sporting code of the amateur

Figure 2.1   The air scout looks skyward. (Reproduced with permission of the Scout Association UK.)

athlete, the courage of the soldier, the gallantry of the medieval knight, the killer instinct of the hunter, and the male bonding of an all-boys school', leaving, 'little room for women' (1994: 282). In Germany the image of the aviation ace was combined with an Aryan myth of a tall, blond and handsome race. As this permeated the Nazi educational syllabus and political propaganda it was sustained by the exploits of the handsome Oswald Boeckle, the national icon who sent the nation into mourning upon the event of his death. In the United States, Charles Lindbergh occupied a similar position as a tall and handsome figure, physically manifesting all that had been romanticized in films, novels and adverts (Pisano and Van der Linden 2002; Wohl 2005). Pilots were often the poster-boys – the characteristics embodied – of various political and national movements (Kennedy 2000).

For the youth of Britain, this idealized aesthetic creature came from a slightly different background to the German one. A Victorian and Edwardian education, along with popular and patriotic boys' magazines such as the *Boy's Own Paper*, *Boy's Own Magazine*, *Magazine for Boys*, *Aunt Judy's Magazine* and *Boys of England*, served to promote an idealized boy body of a middle-class origin (Figure 2.1). Such ideas share their origin in the 'muscular Christianity' developed in Thomas Hughes' and Charles Kingsley's novels, reacting to the poor social conditions of late Victorian Britain, while absorbing the pedagogic teachings of the scientific study of physiology (Haley 1978; Hall 1994; Putney 2001; Vance 1985). Hughes and Kingsley linked strenuous physical exercise to the aims of spiritual development and wellbeing and by so doing designated a body ethic of manliness and action (Bradstock 2000; Dowling 2001; Holden and Ruppel 2003). It was a physically perfect as well as active person who represented the Empire (Springhall et al. 1983).

The aviator's face came to stand for particular moral values and signs of certain qualities of character that could be physiogonomically read (Currell and Cogdell 2006). As Knowles and Malmkjær show, representations of maleness in many Victorian narratives located the *active* and the *physical* in the 'superordinate *face* or its co-metonyms *eyes* or *mouth*' (1996: 89). This may be seen as 'detected evidence of the young hero's superior moral qualities to match his physical attractiveness' (1996: 89). The face could be broken down to show honesty through the eyes or the firmness of the mouth. In Edwardian Britain, these ideas were later combined with a form of Social Darwinism which advocated the notion that fit and efficient citizens would invigorate the health of the Empire (Dobbs 1973; Hynes 1968).[17] Although it was character and qualities the air-minded movements wanted to change, it was the body, '[a]lways and everywhere the body [...] indubitably *there*' (Cunningham 1988: 162–3), which could be read, animated and altered.[18]

Both contexts had several implications for the image of the airman and the social construction of the youth body, creating two predominant ideologies. Dystopic aeronautical novels such as H. G. Wells' *War in the Air* (1908) characterized the neglect of the human physique by tying it to Britain's disregard of the air. Wells writes of the sedentary and grounded Bert Smallways, the book's main character, as 'a vulgar little creature, the sort of pert, limited soul [... who] had lived all his life in narrow streets, and between mean houses he could not look over'. Many other novels described the heroic physique and mentality of the ideal airman, celebrated by Jack Hemings, M. E. Miles and J. F. C. Westerman, among many others (Miles 1936; Westerman 1938). A frank and open face with a strong jaw saturated this imagery.

The image of an aviator 'as a special kind of man, daring, resourceful, and a master of new technology', could be read by the face (Paris 1993: 129). One reporter, stirring every 'mental faculty to penetrate behind the surface of his physical shell' on meeting with Wilbur Wright, subjected the aviator to a physiognomic examination of his 'high-domed forehead' and his nose, which somehow instantly reminded the reader of 'Sherlock Holmes'. With Wilbur's 'close-shut lips' relaxing 'with most extraordinary mobility to join' the eyes in smiling, the whole attitude was 'one of absolute composure'.[19] In Germany, 'the godlike' faces of even women fliers such as Marga von Etzdor were 'a composition of strength, confidence, and unique technical ability', whose nose might bear the crook of 'all born leaders', whose eyes could fix themselves 'like blue darts of flame to the objects of the outside world', and whose characteristically small ears might 'hear even the crystallization of atoms' (cited in Fritzsche 1992: 158). Famous pilots such as Douglas Bader, Cecil Lewis, Alan Cobham and Claude Grahame-White exemplified this persona.

Mobilizing the air-minded body

Tensions lay between these idealized images and the physical educational systems the aerial industries and organizations wanted to pursue. The aesthetic was altered and disassembled as the body-subject was opened up. The author of the Biggles stories, Captain W. E. Johns, saw through the aesthetic and considered that it was a mistake to forget the activity, hard work and endurance necessary to develop the physical character traits so evident in the popular imagery. Johns discussed the myth that physical beauty necessitates bravery and valour in an article for the *Air Training Corps Gazette*. He argued that 'the general impression is that to be tall, strong and handsome is automatically to be fearless'.[20] Johns moved quickly to dispel this myth (a myth he specifically located in National Socialism): 'Looking back over a good many years of flying I cannot recall ever being struck by the association of beauty and sheer bravery.'[21]

Johns went on to write that valour and bravery were not something that could be identified by visual and representational means: 'No one, not even a panel of doctors, can tell, by looking at the outside of a man, what is under the skin.'[22] For Johns, it would take practice and action to determine these qualities, for the 'fellow probably doesn't know himself until he has been in action'.[23] Johns recounts a story of a pilot he used to serve with:

> I remember a certain pilot of 55 Squadron. You will get an idea of what his face was like when I tell you that his nickname was Pug. It wasn't only his face. He stood about four feet nothing and had bow legs. Yet who did the dirty work? Pug. To whom did the C.O. turn when he needed advice? To Pug.[24]

While Pug's physical appearance may not have resembled the stereotypical ace pilot, Johns illustrates how perceptions and surfaces can be deceiving. He conveys the message that bravery and valour are not essential pre-givens but things that are earned and developed through practice. Moreover, the aircraft and the sky are presented as a space that escapes the normal relationships of everyday life which restrict a body like Pug's. For, '[o]n the screen his appearance would probably have been greeted with a yell of mirth, but in the arena of God's blue sky his arrival in a dog-fight was usually the signal for the enemy to retire. If ever a man did not look like a hero, it was Pug.'[25] The organizations re-formed the body of the airman with the traits of hard work, energy and character. Valour and integrity were not predetermined by attractiveness, but were the product of the airman's performative actions and practices.

Notwithstanding all the efforts of Johns and others to decry the existing body aesthetic, the look of the cadet and scout uniform were still vitally important. Uniform had obvious symbolic meanings associated with prestige and bravery (James and Beaumont 1971; Joseph and Alex 1972; Kuper 1973; Mechling 1987). Uniforms for the ATC were designed as a replica of the RAF uniform, whilst the Scouts modified their existing troop design. Emphasis was placed upon the uniform's symbolic value.[26] The proposed uniform for the Corps included a tunic coloured RAF blue with shoulder straps and breast pockets, chromium buttons, and a black leather belt with a chromium buckle. There is no doubt that the aesthetic form and the connotations of the ADCC and ATC uniforms provided one of the biggest attractions to boys at this time. An air cadet in his teens, Alan Sillitoe (1995) described how the uniform was often withheld for three months while the cadets were put on probation to ensure their full commitment and that they were not just joining to take the uniform. For the Air Scouts, the traditional uniform was modified to reflect the aeronautical theme of the branch. Scouts could wear a beret and trousers if senior enough. Badges also embodied performative endeavours of hard work and discipline (Mechling 1984). The first aviation-related scout badge was awarded for completion of the 'Airman's Course for Scouts' organized by the Young Aerial League. The purpose of the badge was given on the lines of a potential aerial invasion.[27] Badges should 'develop the inventiveness and powers of observation of the Scouts'; it could give them 'an intelligent understanding of the aeronautical movement at home and abroad'; and it could, through activities like model building, give scouts a channel to express their aerial ideas.[28]

The badge was the symbolic manifestation of the physical work needed to achieve it, forming an exertion–affect reward economy of performance and display. By the time the Air Scouts were fully formed, three badges had been developed for scouts to attain. The 'Air Apprentice' badge could be realized by the demonstrated knowledge and practice of such skills as the proper way of conducting oneself at an aerodrome, knowing how to be of help to a pilot by indicating wind direction or removing chocks, and making a model aeroplane that could fly 25 yards. The 'Air Mechanic' badge, as it sounds, was more orientated to the workings of the aeroplane, and scouts would need to know the inner apparatus of the internal combustion engine as well as to make competent use of hand tools. Finally, the 'Air Navigator' badge developed a scout's practical knowledge and awareness of topics such as the weather. Scouts were also required to be able to plot on a map a route given by an examiner.

As well as earning the badges and the uniforms, it was proposed that by wearing the uniform one became aerial and was 'instilled with the spirit of the RAF'.[29] Through the act of earning and wearing the uniform the Corps imagined the aerial sensibilities of the RAF could be inscribed upon the cadet-citizen psyche. The wearing of a uniform built a collective feeling of togetherness or belonging. It was just as important for instructors and squadron leaders. Several wrote to the Air League for permission to wear their own badges and uniforms earned in their stints in the RAF. A letter from the governors of King Edward VI School in Nuneaton discussed a situation where the headmaster of the school would not take up leadership of a resident ADCC squadron unless he was permitted to wear his Royal Observer Corps wings.[30] In this instance, uniform was both a sign and a conduit for command and authority; it inspired respect for and gave a 'dignity' to the leader.[31]

Contacting the air

With the opening of the aerial body image to the scrutiny of experts and educators, the young body needed to be made receptive to specific geographical contexts and spaces. Engaging and moving around in a particular context could develop the air-minded boy with certain traits particular to the situation. For the Scouts, these sorts of ideas were extended from Baden-Powell's and Ernest Thomas Seton's naturalist ethos of open-air camping, from which the American frontier, the imperial colony or the raw nature of the African jungle were perceived as sites of nation building, character, manliness and capability (Cronon et al. 1992; Gilcraft 1927, 1932, 1934; Kearns 1984; Matless 1998; MacDonald 1993: 143–4). Spatial exposure could transform.

In part, the scouts' contact was made through the geographical imagination. The story at the campfire could tell of the exploits of aviators in far-off countries. Tales of aerial adventures across the Empire became popular subjects for boys' fiction. Indeed, even RAF cadets were not immune. The RAF's colonial involvement in the Orient was a subject to fill the time in the cadet college's debating society and the pages of its magazine. The 'idea of Arabia' (Satia 2008) was circulated through narratives in a manner that rid Iraq of its romantic Oriental myths. A flying officer's report spoke of finding Basra 'nondescript, dirty, dusty, dreary' with a 'few polluted streams or creeks' along with 'flies, mosquitos, dust and heat'. Baghdad left an even worse impression, being described as 'disillusionment complete and entire. Of romance there is none, or what there is is so deeply hidden that it passes unseen even to the diligent seeker.' The people the officer found were seen as 'lethargic' and the cityscape undrained and filthy. Children had 'sore eyes and blotchy skins' and were 'poorly clad and unwashed'. The young men the officer found were 'un-ambitious, content to spend a youth in a blind alley'.[32]

The dust, haze and sweat of the desert were easily imagined, yet any real physical contact was problematic for the organizations aimed at the younger youths. The Scouts were undecided how they could overcome an idea of aviation as a spectacle. The inability to do aviation-related activities had initially halted the beginnings of a branch of Air Scouts in the 1920s. Ideas first put forward for the branch were resisted because it could not be seen how Air Scouts would be able to *do* something – key to their mantra. Proposals for a branch came from a Scout group in Bournemouth, where a local RAF pilot had offered to begin a troop.[33] The idea was passed on to Baden Baden-Powell, who considered the proposal but could not see how the Scouts could get involved. The Secretary of the Association, Piers Howell, wrote to Lord Hampton, who was championing the cause. Although Howell could see the initial usefulness of such activities, and conceded that for the 'first week or two they will be thrilled by seeing machines on the ground or in flight',[34] he drew upon the familiar Scout derision for spectatorship and passivity: '[M]erely watching a sport has always been discouraged in the movement. [...] They must have something definite to do.'[35]

The Scouts could not see how, given the limited availability of aircraft and aircraft expertise, they could become actively involved.[36] Lord Hampton wrote a letter to the Secretary of State for Air, Sir Sefton Brancker, discussing some of these issues:

> In all our ports and fishing villages you will naturally find a large number of boys who have, as one might say, the sea in their blood. From the earliest age they have gone down to the sea in ships, or at any rate, in small boats, and from constant contact with the sea, they have got 'sea mindedness'. Unless our Air Scouts – should they come into being – have an opportunity of getting air-mindedness in the same way, namely, by fairly constant contact with the things of the air, I don't think we shall get very far.[37]

Lord Hampton followed up their decision with an article in the Scout's *Headquarters Gazette* which stated: 'Air Scouts would have to stop short at theory, except for an occasional joy-ride without the thrill of personal control. They would be lookers-on, and not doers.'[38]

Although the Scouts' access to the air was limited, the ATC, and especially the German youth movements, were able to enjoy significant exposure through what was known as 'air experience'. In fact, the promise of air experience was one of the main ways the youths and cadets were recruited. The Air Ministry had embarked upon a programme of providing financial assistance and subsidies to gliding and light aeroplane clubs to stimulate flyers and pilot training. Moreover, they also subsidized clubs to help train the ATC cadets. Being able to promise cadets the opportunity of flying was seen as the ' "carrot" activity to stimulate recruiting and then to induce air-mindedness'.[39] It was also one of the main reasons recruits were retained. An internal report stated: 'There is no doubt whatsoever that a reasonable amount of flying for ATC cadets is essential if the corps is to be kept alive.'[40]

Only if the movements could offer youths contact with the air would successful recruitment continue.

Gliding patently helped recruit potential cadets. For those who promoted the ATC, the experience of flying was emphasized in two main ways. Firstly, it was described in liberatory metaphors of freedom and emancipation. Resembling the frontier mythologies discussed above, gliding and flying were supposed to provide freedom away from the restrictions of gravity and the ground. Gliding was seen by many aero clubs as a kind of pioneering activity. Those taken up would negotiate new territories of thermals and invisible wind currents; they would actively discover new realms of airspace. From the glider they could experience pure nature distilled from the polluting oppression of urban and terrestrial life (see Fritzsche 1992 on the German example). Secondly, they could discover new and different ways of working together: 'It is a pastime where the claim of the individual must inevitably subordinate itself to the benefit of the whole and therein lies the greater enjoyment.'[41]

For those able to be taken up by a powered aeroplane, a report in the ATC *Gazette* described the initial feeling of complete freedom, even beyond that of the glider. In one's first ascent the youth lost their aerial virginity and transformed into men.

> You realise for the first time that you have an aeroplane with a powerful motor under your full control to go – within limits – where you like, when you like and how you like. I can promise you that it is certainly one of the greatest moments in your life. You will have many other thrilling adventures in the air, but I am sure that all men of the air will remember most vividly their first solo.[42]

Flying also put cadets and youths in direct contact with airspace, providing an escape from the earth, which was mediated by the ever-closer interplay between man and machine (Lewis 2000; Thrift 2000a). Again, the aerial body needed to be exposed to these experiences as soon as was possible. Learning to fly at or about the age of adolescence would grant flying experiences 'with an ease and absence of strain not possible to acquire if flying is learned after complete maturity'. To learn the skills of flying at a younger age would encourage its practice to sink 'below the consciousness and becom[e] a natural and effortless habit', and thus as easy as riding a bike.[43] Giving the cadet air experience at the most 'impressionable age' builds 'confidence that they can handle at least one kind of aircraft. It also has the valuable effect of stimulating his classwork and lending colour and realism to what might otherwise become dull routine' (Taylor 1946: 45; see also Harcombe 1946).

Once the cadet was able to negotiate and react in the air, it was important to see the ground. The new perspective from the sky would figurally change his take on the world. For the young Alan Sillitoe the perspective opened up new opportunities and new knowledges (Sillitoe in Daniels and Rycroft 1993). Similar remarks could be found in the *Gazette*, which reported on how familiar landscapes would magically appear 'the same' and yet 'different':

> The street that you always thought was straight will have a kink in it that you never noticed before, and the church steeple which always in your mind had dominated the other buildings has now become curiously foreshortened. The very colour and lighting of the ground will be quite different and will take on a new grandeur.[44]

An urgent aerial geographical imagination was surfacing.[45] The development of this geographical imagination was fostered by the movements as a way to develop 'air sense'.[46] Cadets were taught to develop and hone the kind of innate senses geographer James Blaut (1997) uncovered within the early child, rejecting Piagetian pessimism. From as early as 2 years old, children showed the remarkable ability to quickly understand, decipher and use aerial images (see Fox 2009). Cadets were trained to learn man-made terrestrial features such as docks, ports, roads and infrastructure – obvious strategic targets for their future action. Furthermore, their geographical education emphasized the relationality of locations.[47] When they learnt the location of places such as Tokyo or Delhi, it was important for the cadets to understand where these locations were in relation to Britain or elsewhere. It was also likely that the information given in maps could be outdated or obsolete. An article in the *Gazette* entitled 'Geography Matters' remarked that 'no one can afford time to acquire redundant knowledge [...] for the present artificial divisions of the world may alter radically before they even graduate' (see Thrift and Dewsbury 2000).[48]

Immersion on the ground

Practical knowledge could be gleaned from the cadets' simple exposure to an RAF camp. The open-air surroundings of the camps were important. A review of RAF Cranwell (Haslam 1982), the main training school of the RAF, described the space as a 'healthy place. Nothing checks the breezes which blow in from the North Sea, while in summer it rejoices in a larger ration of sunshine than most parts of the country receive.'[49] Re-emphasizing anxieties concerning the degenerative qualities of city life, the review went on to observe that at Cranwell pilots looked 'just as one would wish them to look, healthy and athletic, well fed and well trained, drilled and disciplined, keen and earnest, and in particular, happy'.[50]

Extracts from a popular book on RAF life were reprinted in the ATC *Gazette*.[51] In the article the author asks cadets to consider what they have learnt by walking around the RAF camp. The author states, '[T]here is nothing like a new life, new surroundings and the open air as a stimulus to mental activity.'[52] This newness and the changing and dynamics of the youth's surroundings would improve their faculties, as would their ability to remember these changes. According to the author, 'Count up the number of things you have already memorised since you joined up, and you will see that your capacity for learning is really undiminished. You have learnt your station number, the address of your station, the names and faces of a score of your comrades.'[53] Moreover, the cadet will have learnt the spatiality of the site by memorizing the 'geography of [his] station – the way to the canteen, the dining hall and the town'.[54] Cadets didn't need to be lectured on this geography in the classroom because it came through their experience and performance of navigating the camp space. The article went on to impress that a geographical awareness, evolved through practice, would simultaneously improve the cadets' faculties for being educated in other topics:

> [N]ow, if you can learn all these things so easily, you will be able to learn other things – things that now appear difficult, remote, obscure – if they are presented to you clearly and are given time to sink into your mind. And you will be able to learn to do things just as easily as you have learnt to stand to attention.[55]

If the RAF camp could not be reached, then it could be simulated in the Scout or the ATC club hut. For the ATC, these spaces were usually composed of borrowed rooms and make-shift spaces taken up from local authorities and community buildings. Like the Scout camps, which emulated the architecture of the imperial colony or the pioneering cabins of the American frontier in order to 'steer boys towards a measure of wildness and ruggedness' (Barksdae Maynard 1999: 27), these spaces mirrored the RAF camp. Following the simulated practices of much of the Corps, the squadron leaders and instructors were encouraged to design and furnish the club rooms as if they were real service headquarters. Formal labels were given to different rooms according to RAF station service room designations. Therefore many club rooms had an 'orderly room', an office for the 'adjunct', the 'CO' and the 'Sergeants', as well as a 'mess room' and a 'Corporals Room'. The door lintels and skirting boards would also be painted to 'smarten' the spaces up.

The spaces were intended to promote an atmosphere of comradeship and team effort. Instructors were urged to make the club room 'homely'. Instructors could put up photos and drawings, making the club locally distinctive as a place for the cadets to gather and socialize. In all, it was argued that the spaces should be designed to instil a sense of 'club room life' and 'atmosphere', 'so that the real and healthy team spirit of good-will and friendship will be prevalent among the cadets'.[56] The club room was envisaged as a sort of learning space, continuing the idea that learning did not happen individually, but was facilitated by team-work and social learning: 'The clu-broom can be used by the cadets as a place in which to talk shop, to discuss subjects such as in Maths and how best to learn this subject, and to exchange experiences, to talk over difficulties in studying and, best of all, to learn good comradeship.'[57] The club room could even be a means to extend the Corps to wider family members and the support network of the cadets. Many of the squadrons organized open days, inviting their parents and friends along to see what they were up to, while importantly exposing them to the spaces and the activities of the corps: '[C]adets are proud to show what they are doing to their friends, the parents get a better idea of what is happening, and an atmosphere of local pride in the squadron is created.'[58]

The importance of the places of the camp and the cadet hall portray a relatively safe yet practical interiorized geography that was meant to prove beneficial to the cadet's developing body, as if a child growing in the womb. Alternatively, for the first RAF flying officers at Cranwell the *outside* of the Lincolnshire countryside was seen to provide a source of inspiration, con-templation and a rich sense of place. Describing the layered strata of mean-ing to be excavated in the landscape, the spaces around Cranwell appeared as important as the camp itself. 'There is an undercurrent throughout the history of a locality which deeply affects those who are connected with the place,' wrote an article describing the new recruits' impression of Cranwell and the *impression* the locality was to leave on their memories, producing 'the spirit of the place'. Such an intangible undercurrent of *spirit* was seen to emerge from the 'ideals and enthusiasm' of Cranwell's inhabitants.[59] Emplacing the character of the landscape with the spirit and ethos of the nation, Cranwell's locality offered both 'strength and restful relief' and a place full of possibilities.

Taking part in beagling (discussed in more detail below) was to take part in a manner of archaeological discovery where one could recover a sense of the history of the English landscape that was captured in rural Lincolnshire. To move through space was also to move through time, to

> rush through the centuries at incredible speed; you'll dash across a road, and see right and left a wide grassy straightness as far as the eye can see either way. [...] You'll see an aged face of a workman standing as still as a stone as you flit past. One look, and your mind will hark back- hark back to something remote and precious, and well-lost elsewhere.

Importantly, this was to recover a sense of something permanent and deeply elemental. It was 'something that Time hasn't changed and foreign-ers haven't spoilt', but it was, moreover, a wilderness touched only by the wind: 'So that is why Cranwell is where it is, and why we are what we are. The wind!'[60]

The Flesh of the Aerial Youth

It became apparent that other kinds of mobility could help develop the youth body, physically and morally. In their activities, many of Britain's air-minded youth organizations followed the work of the Board of Education, who had laid out prescriptive instructions for games, exercises and activities to develop the nation's young.[61] In 1940 the Board released a circular which advocated the idea that developing bodily fitness would lead to a 'satisfied body and a joyful mind'.[62] It would help the 'building of character', 'developing the whole personality of individual boys and girls to enable them to take their place as full members of the community'.[63] In discussing the ATC, the Board also noted the importance of physical sports and recreation to 'hold' recruits for service, giving them the 'fullest possible opportunities to learn and practice the lessons of initiative, self-giving and self-government'.[64] The ATC would develop better and fuller citizens through their training, 'harnessing that eagerness to the national effort, with advantage both to the young people themselves and to the nation as a whole'.[65] These ideas were not only metaphorical. There are strong resonances between the instructions found in the aerial movements and the scientific theories which had pervaded the education system in the United States by psychologists such as Luther H. Gulick, as Elizabeth Gagen (2004) has so usefully discussed. It was supposed that the training of the body through exertion and physical exercise could help train the character and capacities of an aerial youth.

The practices of the ADCC, the ATC and the Air Scouts owe predominantly to the scientific advances in the disciplines of 'new psychology' and physiology, which began to see the body, mind and spirit in unity. They saw the human being as a whole organism rather than discrete elements existing in isolation. Different parts of the being communicated to each other through flows of circulation. In the Scouts, Baden-Powell's practical education borrowed from contemporary physiologists and social psychologists of the time, whose articles frequented the pages of the in-house magazine, the *Headquarters Gazette*. The work of Lauder Brunton and David Ferrier was published or referred to in the magazine, discussing the workings of the brain and the body. Academics such as Brunton had been instrumental in the furore concerning national efficiency at the turn of the century and argued that mental and intellectual efficiency were tied to physical fitness.[66] They incorporated conceptions of the interdependence between the body and the mind: by training one, the other could benefit.

Such ideas became formalized in the training of RAF recruits. The RAF's Medical Department had begun physical and behavioural testing after the First World War, and the incredible development of physical and psychological research in the RAF's Medical Research Branch attempted to improve and protect the new aerial bodies for the stresses, strains and demands of the aeroplane. The body needed to be made intelligible; it needed to be comprehended as it worked in action – moving through space – in order to bring it to the 'highest state of physical efficiency'.[67] The body was viewed as an assemblage of flows that fostered communication between different vital organs. Tests were modified so that, '[i]nstead of the orthodox examination of heart and lungs with a stethoscope there are carried out practical exercises designed to reveal the strength of the circulation as a whole, and tests the object of which is to determine the responses of the nervous system and the degree of control exercised by that system over the body'.[68] Although we will discuss these examples in much more detail in Chapter Five, the concepts found their way into training through the management of almost all forms of bodily activities, from eating food, to dressing oneself, bathing and cleaning, to physical development. Clearly interpreted as a primary means of military disciplination, or the biopolitical managerialism of a 'logistical life', in Foucault's and Reid's terms (Reid 2006), on the one hand, on the other, we will see that drills were deliberately generative of an excess beyond that of quantifiable docility.

The accepted exercises were codified into the ADCC *Handbook of Rules and Regulations*, effectively the bible for each Squadron, and found within subsequent ATC publications. It was argued that exercises and activities could help improve the physiological communication between the brain and the body. The nerves could be improved and strengthened. According to the ADCC's training manual, 'Each exercise is one of coordination-movements actuated and controlled by the brain. The nerve is the connecting link between brain and muscle, and every time it is brought into use by the brain to work a muscle, it is toned and strengthened.'[69] Through repetition, the pathways between brain and body could be improved further, developing what the manual's authors called the power of 'executive action'. And so by 'performing exercises of coordination brain and muscle are brought into sympathy with one another'; a 'quickness of action and presence of mind' could be developed in the cadet.[70] This kind of physical training was supposed to develop moral character and aptitude to serve the nation.

For instance, it was thought that certain activities could help strengthen the will-power of the cadet, making him more likely to act with bravery and morality. Through regular practice and conditioning, these qualities were developed from the 'necessity of always having to use the brain in the performance of the exercises and in some, to isolate groups of muscle against natural inclination'.[71] Stretching exercises were advocated initially. Stretches for the legs, neck, arms and trunk were all suggested. Some could be done in bed, and others the cadets were told they would have to complete every morning. Even balancing exercises were proposed. While these practices did not require that much physical exertion, they were believed to stimulate the complete coordination between the mind and the body. Exercises like making a cadet close his eyes whilst standing on one leg were devised in order to cause him to make quick and micro-adjustments in his leg to achieve necessary balance. It was thought that this 'fine adjustment between opposing groups of muscles' would 'stimulate the brain and develop the power of thinking quickly and acting immediately – the most valuable qualities in an airman'.[72] These micro-mobilities were later supplanted with more strenuous and bigger bodily exercises for the more advanced cadet.[73]

Suggestions in the *Gazette* reveal the importance ascribed to fitness in the civilian world: 'In civilian life an employer may be impressed by your brisk movement, by your courtesy and discipline before him: another man may not have that.'[74] Air Commodore Chamier went on to urge the necessity of constant swift movement, not only when in exercise: 'You men must learn to move fast: double to your lectures, double back, be brisk. In the air things happen quickly, and a fast-moving body will develop a fast-thinking mind.'[75] Indeed, Major Gorewood wrote to the RAF cadets that the immobile activity of bathing should be definitely avoided as it threatened to pacify the body into contemplative silence. 'Don't sit for hours in a hot bath reading a novel or studying form,' he wrote, 'A quick tepid bath daily is sufficient. Avoid cold baths. Stimulate the body by a thorough rub with a rough towel.'[76]

Such training was also tested and examined as part of the cadets' 1st Class Cadet Test and the Proficiency Examination for the potential RAF recruits.[77] The marks of such tests would be posted on a board in the club room to inspire individual competition between cadets, and induce them to improve their performances. Tests included those that measured agility by performing a 100 yards sprint, high jump, long jump or vaulting. Strength was examined by pull-ups, press-ups, forward heaves and rope climbing, while endurance and muscular coordination were tested by running and walking, swimming, throwing and catching and quoits.

Drilling was impressed upon the cadets through a tailored guide which was borrowed from the RAF's own manual. According to the instructions, the purposes of drilling and other training exercises was to 'develop certain attributes, both moral and physical, in the officers and their boys'.[78] It was believed that the physical exercises would lead to the natural development of character. Drill specified youth bodily motions in minute detail. The drills were to be learnt through a combination of performance and observation. First, the instructor would explain the drill and how it was to be accomplished. He would then carry out the drill by performing it himself. The squadron were asked to copy his actions, but to do so in slow time so that he could witness and understand their movements. The instructor would then monitor the squad's progress before correcting mistakes and inconsistencies. The cadets were supposed to repeat the drills over and over again so that they were remembered. After improvement was made in the exercise, the cadets could then move in 'quick time'.

Drilling was used for many public displays to develop pride and enthusiasm in the cadets and in the nation, in Britain and elsewhere. Rallies were held on a much larger scale, bringing together cadets from squadrons in London, Birmingham and Manchester. Notable occasions included the united rally of youth movements which took place in Hyde Park on Sunday, 27 September 1942 and the Lord Mayor's show, also in London, in November that year. These displays were often flag-waving and patriotic occasions and were a 'picturesque and inspiring' sight.[79] The instillation of not only air-mindedness but also patriotism and good citizenship was supposed to be generated through the spectacle of movement. The youth moved with bravery and briskness, and, according to *The Times*, dressed in their 'light shirts with bare arms', they showed 'clear evidence of toughness'.[80]

Yet the ultimate goal of the embodied performance of drilling was not spectacle, but so that the cadets could develop certain qualities of an aerial character. By moving in rhythmic unity, 'a primitive and very powerful social bond' would well up among them. Moving in this way created 'an intense fellow feeling' or a 'primitive reserve of sociality' (McNeill 1995). For example, it was thought that being able to follow drill instructions without question would instil 'the habit of instinctive obedience which in the swiftness of the air action and stress of battle *must* be instinctive'.[81] On marching in double time, the ATC's drill book states:

> Each boy will step off with the left foot about double on the toes with easy, swinging strides, inkling the body forward but maintaining its correct carriage. The feet will be raised cleanly from the ground at each pace but they must not be lifted towards the seat. The thigh, knee and ankle joints should all work freely and without stiffness. The whole body will be carried forward, without increased effort, by a thrust from the rear foot, which will then be moved straight to the front and the toes placed lightly on the ground. The arms must swing easily from the shoulders and sufficiently clear of the body to allow of full freedom for the chest. They should be bent at the elbow, the fore arm forming an angle of about 135 degrees with the upper arm (mid-way between a straight arm and a right angle at the elbow) with the fists clenched and the backs of the hands outwards. The shoulders should be kept steady and square to the front and the head erect.[82]

A descendant of the disciplining techniques of army drill analysed by Foucault (1977), these movements were meant to be copied exactly, positions and postures measured in degrees and inches. It was vitally important that the drills and other games and exercises be performed correctly, for if carried out carelessly they could 'do more harm than good'.[83] It is in this sense that the right movements were vitally important to the conditioning of the *right* kind of youth.

It was impressed upon each cadet that these individual motions, which instilled the character traits of 'obedience, steadiness and self-reliance', making 'him alert and giv[ing] him a smart bearing', could then be combined with other formations and choreographies as he learnt how to 'wheel', to 'form-squad on the left', 'on the right', to 'about turn', and many others. This would give him 'pride in himself' while also having the effect of stimulating pride in his squad or flight. And as the movement of a flight could then be combined into the movement of a squadron, and then a number of squadrons so that they could move as one formation, together this would spread not only the standard of individual discipline but a 'mutual confidence between all ranks in a unit'.[84]

Through these incredibly complicated and detailed instructions, there was a sense that the cadets needed to first learn mastery and awareness of themselves, before it could be extended outwards towards their troop, squadron and eventually the nation. An upscaling of unconscious awareness was to emerge by moving together in time. As control over the most minute gestures of the body and the wider awareness of the other cadets improved, it was suggested that this could progress the cadets' awareness of their fellow men, wider society and their standing in the development of the nation. The drills could inculcate '[p]atriotism and esprit de corps and the moral qualities on which they are founded: pride of race, loyalty and a sense of honour', among others.[85]

Simulation

Given all the activities the ATC and the Air Scouts were able to accomplish, much of their training occurred through mimicry and imitation. The movements, in their physical and kinaesthetic approach to learning and educating youths in air-mindedness, developed techniques that employed literal performances of mobility. Performances in organizations such as the Scouts are, of course, not unheard of, with the tradition of Ralph Reader's Gang Show, which became an incredibly popular display put on by Scouts and RAF personnel.[86] But as the public performances of the Gang Show were used to entertain wider audiences, and in some cases install a sense of patriotism and nationalism in those who watched, it was in the specific performative actions used to construct displays that youths were made aerial and better citizens. These were not necessarily based on physiological grounds by the conditioning of the muscular consciousness, as we have been discussing. Acting air-minded meant specifically chosen mobilities and activities which mimicked the actions of their 'real' counterparts.

Manliness and camaraderie

Sport and games were one of the most important training tools for the youth's body, personality and character. The use of sport as an allegory for flight was common, especially by those describing the dogfights of the First World War as a form of chivalric duty. Books such as Cecil Lewis' *Sagittarius Rising* (1936) highlighted the way flying was considered as a form of a medieval tournament. Michael Paris shows how these ideas also pervaded cinema. Films such as *Hell's Angels*, *The Eagle and the Hawk*, *Dawn Patrol* and *The Grand Illusion* romanticize the aviators' world as one of devotion to duty, where men follow a 'strict warrior code' (Goldstein 1986; Paris 1995; Pendo 1985). Camaraderie united airmen together, whether friend or foe, where men jousted hard on the pitch but relaxed afterwards as friends: '[F]lyers fought with the ardour of schoolboys flinging themselves into a football scrimmage' (Morrow 2003: 176). The pilot continued the 'games ethic' so present in Victorian and Edwardian public school education, the background from which many of the pilots would have come (Kirk and Twigg 1995; Mangan 1998, 2000).

Although the idea of the flying ace as chivalric sportsman may have deteriorated in the years of aerial bombardment during the Second World War, the youth movements took up the allegory in training. Part of their purpose was physical development. Games were used as exercises to construct the type of fit and healthy aerial youth we have been discussing (see also Walford 1969). But their other importance was to resemble and emulate the sort of experiences cadets and scouts could encounter in the air. For the first RAF cadets at Cranwell, sport and drilling were supplemented with aristocratic 'blood sports' that followed the experiences of the Army. It was thought that their experiential training could be applied just as easily to the Air Force. A senior officer of RAF Cranwell who was a particularly keen hunter encouraged the new cadets to spend their mornings hunting foxes with the station's resident beagle hounds. Hunting was believed to enable the training of cadet 'initiative, quickness to appreciate a situation, and physical endurance'. All of these developments were seen to be easily accessed and 'readily obtainable [...] in the chase'.[87] A sport such as rifle shooting, once learnt, would lend itself to producing a 'better all-round man', cultivating the virtues of self-control, determination and concentration, along with patience and clean living.[88]

Reports described how hunting entrained the body's instincts in a way that was stimulating and joyful. One article emphasized the 'the *drive*' and 'the pace which you must have to kill a fox'.[89] The uncompromising and uncontemplated 'thereness' of the drive constituted the joy and the benefit of the hunt. In the same issue of the magazine the art of hawking was discussed as suggestive of the animalistic joy of flying and the satisfaction of the kill.

Now for the first time you are to see for yourself the 'speed of a hawk'. He passes like a shooting star right across the sky, and is soon almost over the speeding covey. There is a yet grander sight to come, for after a few more mighty strokes his wings are shut close and he hurls himself, with truly appalling speed, down through the sunlit air. The most wonderful sight in nature, this 'stoop' of the Peregrine on its quarry must be seen to be believed. The eye can scarcely follow it, but the ear can hear the high scream of the bell and the hiss of the hawk's wings as he goes.[90]

Juxtaposing the activities of the hawk with the cadet pilots, the article evokes the idea of flight as an inhuman capacity to be treated with awe and satisfaction:[91] 'Diving upon his quarry the hawk attacks a partridge with a great "Whack!"', closing his wings around it and leaving a 'little puff of feathers hanging in the sun'.[92]

Boxing was also used as a popular tool to teach cadets how to fight. In a series of articles entitled 'The Young Airman', *The Times* stated: 'It must be remembered that many of the boys had never struck a blow in anger'; the reporter later overhearing the boxing instructor yell: 'You will be out fighting the Hun soon, and if you can't fight, well, 'e gets you.'[93] As Mangan argues, this 'games cult' helped to project 'an image of boyhood and subsequent manhood' (1998: 191). This embraced 'essential virtues, which contained the seeds of imperial conquest and consolidation', and, in the ATC's case, the means to master the air and defeat Britain's enemies. The threat of the bomber's 'knock-out blow' (see Chapter Six) shadowed the bodily performance of the cadet's jab, cross and upper cut.

The emphasis of competitive sports was tempered with the use of team games such as football, rugby and many others for their powers to inculcate a spirit of comradeship, team-work and also discipline. For instance, football could be seen to quite closely resemble the dynamics of the aerial battle. The interrelationships in a football team were argued to be remarkably similar to a squadron. It was suggested that knowing one's place on a football side, knowing one's jobs as a sweeper, defender, striker or midfielder, would be analogous to knowing one's role in a squadron or in an air crew. As an article by Lord Wakefield in the *Gazette* put it:

> If a team is to be successful, each individual must know what is expected of him, and he must know how to help a team mate who may be hard pressed. The right team spirit and good team work naturally follow when members of a team thoroughly understand what they are required to do, and how each can best help the other. [...] In every hard and equally contested game, a time comes when the slightest easing up spells failure. Success can only come to the fittest and best trained side. This means that each man must be fit or he will let his side down. All this, which is so apparent in a game of football is equally applicable in the air. [...] Testing times comes as in a game of football when things go wrong. If each man knows his job and all work under the captain as a team, then as we have seen time and again in this war, in spite of enormous difficulties, a mission is successfully accomplished and the aircraft and crew get back home.[94]

Even synchronized swimming was regarded in a similar light as it was expected that cadets would gain a greater awareness of their position in relation to their fellow cadets.[95]

Youths could develop the qualities of aerial living through the more literal modes of performative learning to be found in games. For the Scouts these techniques carried through from Robert Baden-Powell, who, Elleke Boehmer notes, 'viewed all social roles and responsibilities as manifesting themselves at the level of performance' (2005: xxix), borrowing from the burgeoning field of physical learning techniques. In the example of the game 'Crash', featured in the Air Scout handbook as a suggested exercise, the object of the game is to simulate an accident in which the pilot and an observer of the plane need rescuing.[96] Scouts must pretend they are the fire crew and the plane is represented by a number of chairs laid down on the floor. The participants should imagine the intense heat felt in such an event and not venture nearer than four yards. While ropes and staves are given to the scouts, any other instruction is to be withheld, leaving 'the rest to the ingenuity of the competitors'.[97]

Other games were utilized to develop the scouts' strategic sense of air warfare. For instance, one such game, entitled 'Operation X', saw scouts tying wool round their arms. Sticks with black bands painted across them representing targets of primary importance, and white paper sticks to represent secondary targets, were placed on the battlefield.[98] Scouts became aircraft positioned by their 'fighter chief' and 'bomber chief', who would perform the role of their real equivalents by practising the strategies and tactics of this form of air combat. They would learn how to take defensive and offensive manoeuvres, how to send out reconnaissance aircraft, and they discovered ways to make surprise attacks, decoys and bluffs. In all, this game was intended to educate the scouts on the value of strategic warfare, where planning could be made to be 'fun'.

An element of the intangible 'fun' was vitally important to the success of the schemes. In 'Fighters and Bombers' scouts physically became and embodied the fighter or bomber aircraft through their movements. The troop gathered into a formation of an air squadron and move according to the instructions of the Squadron Leader(as performed by the Scouter). The Scout troop's performances of the various aircraft sounds made the activity a particularly noisy one. Once the order 'Take off' was called the following movements were practised: '[T]he circle begins to creep round, then "fly at 250" and the circle begins to walk, then "Open throttles" and the circle runs (this can be varied from then on, but instead of "Take off" the order "Steady glide" is used instead).'[99] Once the leader shouted one of the scout's names, this scout had to fly as quickly as he could into the safety zone, a circle of about two feet in diameter drawn somewhere on the field, with the rest of the squadron rushing to descend on the scout in an attempt to 'shoot him down'. Scouts became the aircraft and their bullets, accompanied by the deafening screaming of scouts performing the diving fighter's special effects.

In the ATC, similar strategies were adopted and practised throughout the squadrons. Bicycles became used as substitutes for aeroplanes. Wings were attached to the handlebars along with a compass. Squadrons would practise formation flying on their bikes, as if they were in a real fighter squadron. The games would help to instil the sense of leadership, organization and awareness. 'Aerodrome Attack', practised again on bicycles, was played out by cadets with silhouettes of aircraft attached to their bikes. Two sides were set. One side would defend and the other would attack. Airborne cadets (on bikes) could approach their targets by whatever route. In the event that cadets should meet, they would swap silhouettes and say, 'name of aircraft please'.[100] If one cadet should fail to know the name of the cadet's aircraft, and that cadet were to answer correctly, the incorrect cadet would then relinquish his silhouette, having been 'shot down'. The game encouraged the cadets to think up strategies of concealment and approach at the same time, testing their knowledge of air-craft and field-craft, while also 'compelling them to use their initiative, powers of observation and pluck'.[101]

The mobilities resemble the kinds of choreography futurist F. T. Marinetti prescribed in his manifesto for *The Futurist Dance*, subtitled *Dance of the Shrapnel, Dance of the Machine-Gun, Dance of the Aviator*, and were represented in Gianni Censi's dance-piece *Aerodanze* in Milan in 1931. According to Anja Klock, each female dancer in Censi's piece was to perform the movement of an aeroplane as follows:

> [S]he should draw, with her opened arms, the long hissing curve of the shrapnel that passes over the combatant's head, showing how it explodes high above or behind him; she should produce the hammering sound of the machine-gun with her feet; she should wear a large celluloid propeller on her chest that vibrates with every movement of her body. (Klock 1999: 399)

The role-playing games performed by the youths were simulacra – the physical embodiment of the movement and behaviour of their real counterparts – whether they were pilot, or aircraft, or fire truck or bullet. The intention in this context was not to internalize the technological apparatus, or to externalize them through their representational practice. Instead the emphasis was on developing the body's faculties through its movement and exertion. The youths literally acted out and became aerial through their mobility, they developed their character, and they learnt the bodily skills demanded (Matless 1998: 89).

Model airmen

In the ATC, model building of aircraft developed as another aviation-minded practice and can be situated in the wider use of modelling activities in many British schools during this time.[102] As geographer Teresa Ploszajska (1996) has demonstrated, model building became an important means of teaching an especially geographical education. Models of hills, mountains and seas were constructed by pupils, requesting their physical and active involvement in assembly, thereby awakening their tactile and haptic senses (Buhler 1933). In the same way, I will argue that the youth's construction of models was designed to be constructive of them. Models could help construct *model* aerial subjects.

As early as 1926 the Air League had discussed the importance of model building and model-building organisations such as the Society of Model Aeronautical Engineers. In an article in the *Air League Bulletin* it was written that model flying 'is undoubtedly a highly educative means of interesting the people of this country'.[103] The Air League discussed the importance of using model flying to arouse interest in the public. As was the case with the airshows, the public could not merely be told about model flying, instead they had to witness and experience its performance.[104] And yet, the public demonstration of model aeroplanes was by far the least important element emphasized in the League's encouragement of air-mindednesss. Instead, it was thought that through the model's assembly and construction specific educational values could be nurtured.

The building of model aeroplanes did not become part of the academic syllabus in schools as it did in the United States and Germany. As Joseph Corn (1983) shows, educators in the United States 'infiltrated' the syllabus of Maths, History and English lessons with an air-minded theme. This did not mean, however, that an appetite for model building was lacking in Britain. One pamphlet the Air League produced received 800 orders from headmasters and technology tutors around the country for the detailed aeroplane designs contained within it. Outside of schools hundreds of clubs sprung up around the country as a way to stimulate this activity, whilst also catering for it. The Skybird modelling kits were produced on a mass scale in conjunction with a league which was formed to oversee the accelerated progression of the model clubs, becoming known as the Skybird League. In collaboration with the Air League, the Skybirds began their own publication, *The Skybird*, in 1933, before forming the popular *Aero-modeller* in 1935.

For the Skybirds, modelling was a kind of pioneering activity of considerable national importance. Those who joined the League were 'helping to encourage airmindedness in this country just at a critical period when, as a nation once holding a premium position in aviation, we had fallen to fifth place. No other country has yet produced any scheme quite like SKYBIRDS.'[105] By 1935, some 300 clubs had been set up around the country. Along with the classes run by the Society of Model Aeronautical Engineers to teach people how to build model aircraft, the Air League described such schemes to be of 'national importance, insomuch as it helps to spread an "air sense" amongst the people and is very often the means, especially in the younger generation, of producing the potential airman of the future'.[106]

In the Scouts, as we noted above, as part of the Air Apprentice badge, scouts had to be able to build a model aeroplane which would fly at least 25

Figure 2.2   Air scouts fly their model glider. (Reproduced with permission of the Scout Association UK.)

yards. While building and flying the aircraft would give the youth an intuitive feel for how things fly, the purpose of the test was to also develop other qualities. To see something materialize from idea to tangible thing, from start to finish, was powerful. Moreover, '[o]pportunities of ingenuity and individuality' were afforded by such practices.[107] The practice was also beneficial for the youth's body as the micro-bodily movements of 'manual dexterity' and the mental faculty of 'patience' were viewed as necessary for the successful completion of such a project.[108] In other words, a combination of physical and intellectual qualities would be tested and developed by the practice; the 'handiness' of the scouts was a primary objective. In other venues, modelling was even perceived to be healthy. Through the outdoor exertion necessary to launch the glider, model gliding could improve the channels between mind and body. For the Secretary of the Model Aviation Club: 'Model aviation has been described as one of the most satisfying of hobbies. This it undoubtedly is, giving fascinating employment to brain and hands, and healthy outdoor exercise' (see Figure 2.2).[109]

Probably the most important element in modelling was realism. Magazines, journals and instruction books were filled with information on the most minute details of planes so that they might be copied. *The Skybird* often featured large descriptions of the aircraft detailing what they were like to fly, how they were made, what they were made of, and, importantly, some of the feats they had achieved in the First World War, or perhaps in setting records or winning races. By building the models, youths could almost taste the real thing. In one article, *The Skybird* even suggested that veteran pilots, whose flying days were now over could relive their experiences through model making. The article stated: 'Many an airman who flew in the Great War has recaptured the thrill of those dim, hectic years with the aid of SKYBIRDS. [...] Let the "old boys" look back; let the younger boys look ahead.'[110]

The search for realistic models meant that many aero-modellers and Skybirds wanted to try to compose scenes for the aircraft. Adding context to the model planes could seemingly add depth and a sense of reality. Thus, Skybirds brought out and sold thousands of miniature aerodromes, hangars, engine shops, workshops and model airports to house their models. People could buy and make tiny figures to populate the scenes, creating bustling miniatures of a working airport.[111] People played with the scenes in a way they might with miniature soldiers.[112] These scenes would then be photographed and sent in to the popular magazines or to competitions.

Modellers created simulations of their model aircraft in flight. Painted backdrops of the sky could be purchased on which model aircraft could be fastened. Cliffs, mountains, rolling hills and other scenes could be provided as suitably picturesque backgrounds. Or they could be played with. Youths could pretend that their aeroplane moving across the backdrop was a real aircraft screaming across the clear skies. Some modellers used techniques like adding a wingtip of a plane into the scene. According to one modeller, this would mean that 'the idea is immediately given that the photograph was actually taken from a plane'.[113] If placed around three or four inches in front of the line of the models, a realistic impression could be given – disembodying the position of the aerial body from the ground and into the air. Cotton wool could also be applied to the background to make the 'most convincing' cloud shapes. Through all these activities, certain qualities would be inculcated in the youth, whose efforts in 'producing realism by getting the right scale for all components, and natural "lighting" effects' would 'depend entirely upon his ability and perseverance'.[114]

Conclusion

To become aerial one could not watch from the sidelines. Instead the emphasis was on the action and active participation given by the motion of the young mobile body. Aerial life was not inspired through the symbolic representational practices of the aviation spectacle, or the specified practices of observing, but it came about through the practices that composed these representations; through the mobilities that made the parades, displays and films. It was a social, naturalist, physio-psychological and simulational activity, instilled through a multiplicity of mobilities which were concurrently performed and representational (Cresswell 2006b).

The benefits of these mobilities were up-scaled to the nation. Exercises and balancing could help develop will-power. Drilling could encourage the youths' awareness of themselves within their squadron and their country. But most inventive of all the activities the youth organizations developed was to allow them to become the airman through performances. The cadets and the scouts, through role play and games, literally acted out what it was to fly, what it was to plan a battle, what it needed to save someone from a crash. These carefully choreographed practices mixed physical mobility with an imagined one. The games relied as much upon the practical movement of youths running around like an aeroplane as they did upon the imagination to believe they were one.

What we see, then, are the precursors of the aerial subject produced as a particular kind of body: a body readied for performance, prepared for war; a body militarized and poised to step into action; a citizen-body, attaining stronger links with the body of the nation. In the following chapter we see how these choreographies of the body are paralleled by wider mobilities, mobilities of populations and aircraft choreographing the performance of national space.

Chapter Three

The Projection and Performance of Airspace

It is this complexity of and within the airport, its bringing together of so many of the key themes of modern life – mobility and surveillance, consumption and diversity, sovereignty and vulnerability – that makes it a rich metaphor for the larger political community. It is a space distinct from everyday life that reflects and comments on the broader spaces of modern life.

(Feldman 2007: 48)

Introduction

In May 1909 a mysterious shape was seen in the skies above Britain. The shape both intrigued and terrified. Reported by daily and local newspapers, the strange cigar-like object emerged from clouds scattered against the black night. The first reports came in from Suffolk: a skipper of a fishing boat reported what he described as 'a large star rising out of the water' off the coast of Lowestoft. The 'airship' appeared to be sausage-shaped and carried a single light; it resembled a similar object the skipper had seen the year before. Across Britain several other sightings were reported. Police Constable Kettle in Peterborough took down accounts which were later noted by Charles Fort (1974) – father of Fortean science (Dixon 2007) – when investigating the events which became known as the 'airship scares'.

The scares originated along Britain's eastern coastline. In Ipswich at the end of April, a bright light was observed by a group of people and reported by Police Constable Arthur Hudson:

It appeared to be at a great height, and I lost sight of it at intervals. Whilst I was watching the light I suddenly observed a dark object which appeared to be about a hundred yards from the lighted one. I examined it through a pair of opera glasses, and the dark object appeared to be like an ordinary balloon.

After hovering about for a time it passed out of sight in a north-westerly direction. (cited in Clarke 2000)

By late May the sights had spread. In Cardiff, over several days many airship sightings were reported. In the docks the odd outline of a torpedo-shaped airship was witnessed. It was 'very long', imposing in its size, and – frighteningly – it sat almost a quarter of a mile above the houses and trees. Hanging in the sky, it 'remained immovable for a few minutes'. The airship made no noise. Its soundlessness was put down to its distance away from the spectators, which was closed by the aircraft's searchlights, piercing the dark and scanning the ground and sky. There was no 'question about the reality of the mysterious airship', according to a traffic foreman, Mr Harwood. For another spectator, 'The night airship is a fact.'[1] An editorial in *The South Wales Echo* which pondered 'why the east coast should have the exclusive possession of a mysterious airship of the night' was satisfied, having now received various reports in the Cardiff area, that Wales had 'at last seen a "dark object in the sky"'.[2]

In fact the scareships were quite impossible. Experts at the time argued that the technology did not yet exist. Post-war investigations of the apparitions (discussed later) show this to be true. But the 'ghosts' still spread. Almost every coastal town began reporting similar discoveries. People worried about eyes peering down upon them from above. They speculated about the astronauts or alien nationals who had punctured the country's borders. And they feared the contents of the airships' holds. With the country suddenly touchable from across the Channel, the airships expressed and fomented Britain's paranoia that its island security was to be soon at an end (Gollin 1989).

In Jung's (1979) post-war investigations into Unidentified Flying Objects and other 'things seen in the sky', spectral phenomena such as the scareships emerge from both personal and collective contexts of anxiety and tension that are projected onto the susceptible psyche, and onto the sky (see Holloway and Kneale 2008; Luckhurst 2002). Jung argues that flying objects emerge from the 'background' world of the affective unconscious and surface in imagined phenomena. UFOs must have 'a *psychic cause*', he exclaims. The phenomenon must be premised upon an 'omnipresent emotional foundation', a kind of '*emotional tension*' that is to be found in 'a situation of collective distress or danger, or in a vital psychic need' (1979: 8; see also Grove 1970). Cigar- and tube-shaped objects such as zeppelins, as Jung shows, have often featured in witness reports and representations over the past 500 years.[3] The shape, reminiscent of the outline of a spear, 'seems to represent the masculine element, especially in its "penetrating" capacity' (Jung 1979: 131). In their ability to pierce the norms of space and time, the airships performed a relation of projection itself. Bestowed upon the state was the ability to propel – to perform both affective and geopolitical projection and penetration.[4]

Alison Williams' (2007) work in this area suggests that 'power projection' marks one of the most recent uses of airpower. 'Power projection', she argues, 'relies on a number of factors to be effective; some are technological, others are psychological. Both types are needed to gain and maintain control over a specific territory' (Williams 2007: 514). The comparisons with the scareships are clear, both capturing the uneasy relation between the aeroplane, territoriality and the nation-state. The imagined zeppelins, and the real ones that caused such damage in London during the First World War, threaten and menace the integrity of the nation's security – no longer inviolable by attack from above. As *The South Wales Echo* put it in grandiose terms (see discussion in Chapter Six), the aircraft heralded the age of the *air-quake*:

> Civilisation is to be undone and all our methods upset. The age of iron will be over, and the air-quake begun. [...] Humanity will learn how to control and deal with the 'nightmare of the future' when it will be possible to lay Paris, Berlin or London in a heap of smoking ruins before breakfast. We will not pause in our daily pursuits, our commerce, or the preparations for the Whitsuntide and the summer holidays in trembling anticipation and waiting for the era of the air-quake.[5]

In the context of 1909, the scareships tied the zeppelins and all future aerial craft to the nation-state and its boundaries. To whom the air-craft belonged was a key issue in identifying the apparition's sovereignty and, thus, the intentions of that state. As we saw in Cardiff, even the geographical position of where the airships took their shape was an issue of national and regional construction. To host an airship visitation was to be significant.

British military experts and leading proponents of the aeroplane, whilst cynical of the airships' actual existence – parliamentary questions provoked both laughter and silence in the House of Commons – maintained that their mere possibility justified the direction of resources towards the research and development of aerial technologies (Edgarton 1991). The airships could continually threaten Britain's borders by their ghostly arrival, departure and persistence in the air, while, paradoxically, the technology could simultaneously maintain the country's boundaries, making the air enforceable and delimited.

This chapter is concerned with the projective shape of the aeroplane's spaces. Understanding the aeroplane as both architect and janitor of national boundaries with the power to maintain and ultimately destroy the volumes it creates, the chapter explores the spaces of the aeroplane's national performances, from the ephemeral choreographies of the airshow and the social construction of the air, to the hard materialities of the airport terminal.

Before the first official designations or material constructions that we might determine as air-spaces (such as airports or the techno-legal constructions above us [Pascoe 2001]) were semi-temporary events where the first aircraft were demonstrated and shown to the masses. The airfields of Reims and Hendon emerged as key meeting posts for the promulgation of flight's message, the performance of a political community and the projection of the nation into the sky. This politics was aesthetic, made through the art of the spectacular based upon a new technological modernism and various modes of voyeurism, comfort and spectatorship. The affective and ephemeral theatricality of the shows, where pilots displayed their skills and air forces flaunted their prowess of the skies, serve as apt precursors to the *performance* of national space – built, eroded and transformed by the aeroplane.

Building a Political Space: Identity, Boundedness and the Sanctity of Territory

On the 27 April 2009, President Barack Obama's Air Force One, a Boeing aircraft 747, buzzed the skyline of New York City in a terribly misconceived display of Presidential grandeur. The 'photo-op' caused Manhattan citizens to cower in terror, believing a repeat of September 11, 2001. Phone cameras and news footage captured the aircraft flying against the Manhattan skyline, followed by an F-16 plane. The image ghoulishly echoed the famous composition of aircraft-on-skyscraper that has now been reproduced all over the world. People were seen diving into shelter and scattering from the streets, transfixed in anticipation of the plane's possible trajectory of destruction (Sulzberger and Wald 2009). The city's roads teemed with people who had been evacuated from nearby office buildings. A White House official, the Military Office Director who organized the event, duly resigned (Zeleny 2009).

Fly-pasts such as Air Force One's are not that unusual. Of course, they normally happen above less anxious spaces which are not as scarred with the same collective memories as Manhattan. Air displays are still popular in conjuring support for our armed forces, or for remembering the lives and toil of family members (Demetz 2002; Ogilvy 1984). They are public rituals of national identity, remembering and celebrations of belonging (Edensor 2002). Although they are quite unusual in relation to our own positioning to air travel today, they didn't used to be this way.

In Britain, Hendon Aerodrome, which was the home of the RAF's training centre, became the central contact point between the population and flight. Seeing the potential for aerial displays to get people interested in aviation, Claude Grahame-White, the aviator who was to wow city-goers crowding the streets in Washington, led the organization of the first of the Hendon airshows in May 1911. These took on the dimensions of both an educational experience and an awakening that could break through apathy towards the aeroplane. The motto of Grahame-White's company was 'wake-up England'.[6] Without patronizing people, an airshow could prove a step beyond simple education. Grahame-White thought the public resentful of such tiresome pedagogy. 'As a rule,' he wrote, 'the public do not like being educated. When you talk about "educating" grown-up people, it is an awkward word to use.' A better plan was to 'try to interest people, to amuse them as well if you can. If you begin in this way you can often teach them a little at the same time.'[7]

Choreographies of nationhood

The initial summer displays at Hendon provided both an object of interest as well as a warning. '[B]y showing people what the aeroplane can do, thus teaching them the danger of neglecting aerial armament',[8] the displays were intended to condition the display-goers with their clock-work regularity, like a 'drop of water wearing away the stone', as Grahame-White put it. He continued: 'If you are going to make any definite impression upon the public mind with flying displays – as with anything else – you must hold them regularly.' A short and sudden shock or jolt was not the medicine; rather, '[p]eople must get used to an idea. They must get to like it by degrees. So we have been running our meetings very carefully from week to week – making our appeal to the public progressive so to speak.'[9] And thus Grahame-White hoped to arouse an interest 'not fickle or short-lived but sound and enduring'. The aim was to make the aeroplane easily recognizable, to know its capacities.

Those in the aviation industry saw the aerial shows as a persuasive tool of propaganda (Edgerton 1991; Myerscough 1985). For *Flight Magazine*, the success of the aerial derbies lay in their possession of a 'direct appeal to the man in the street'. 'This is an age of propaganda', the article stated. Compared to all the popular events of Lindbergh and Cobham, '[t]he Aerial Derby is in a different class'. People, the article argued, 'talk about it for days before-hand and for days afterwards, and thus a lasting impression of the safety and speed of flight is left on their minds. [...] There is no finer propaganda than this.'[10] Effective propaganda had to produce something far more lasting than the immediate speed and power presented by the machines.

But there was also something about the aerial derby that made it special. It had 'the advantage over most forms of aerial racing by being spectacular (outside the mere commonplace of flying nowadays), free from undue delays and waits', it was supposedly 'intelligible to the spectators, and easily followed at the various points'. A letter to Lord Trenchard described it as a 'wonderful spectacle, and a mighty demonstration of skill, courage and endurance' which could instil a mobility of mind, to 'move to admiration the minds of the 100,000 persons who saw it'.[11] The airshow could do the work of a great and charismatic leader: it could inspire, awe, and, metaphorically, uplift.

The growing crowds, Grahame-White reflected, had gone from apathy towards flight 'almost without realizing it themselves – until they are now quite keen in following all that takes place in the Aerodrome'.[12] A subliminal persuader, the airshows would create a public susceptible to their message, adjusted to an opinion, and who could be urged to believe, for example, 'that sufficient army airmen shall be trained to fly without further delay'. Preferring actions over words, Grahame-White believed that the displays were becoming successful in 'awakening slowly' the people of Britain 'to an appreciation of the aerial menace, and that soon there will be little need for us to cry out to them to act'.[13]

As well as inspiring the public in flight, the airshows were intended to have an 'intrinsic value, apart altogether from their spectacular nature'. The displays of manoeuvres and attacks would 'show the people, in the simplest and yet most effective way, just what flying actually means'. The displays were also training activities. Confidence was a sense apparently imbued in the pilot by the accomplishment of tricks such as a loop. 'Once he had become adept at the art, it may be safely said that not the most violent wind, the fiercest gust, or most turbulent eddy has any errors for him any longer or need cause him the slightest concern.'[14] The pilot's mastery, honed through aerobatics, would, some argued, be essential for the tasks of a 'fighting pilot',[15] though this was the subject of dispute. Small wars occurred between the show's organizers as to whether the displays helped training very much at all. By 1922, the regular Hendon displays were under threat because of their lack of a 'very definite training value'.[16] The issue was the payoff between the spectacular elements of the display and the actual use for training purposes. The pageants were perceived as almost 'inconvenient' and of questionable value given the amount of time and effort that was spent on them.[17] It was proposed that the display be converted into a kind of review shut off from the public's gaze. 'Such an arrangement', it was explained, 'would, whilst retaining the valuable features of the existing Display as a means of setting a high standard of flying, avoid the undesirable diversion of effort on spectacular items possessing little training value to the Force.'[18]

The argument didn't hold much water. The training element of the shows was possibly more relevant to the public crowds than to the aircraft personnel. Mock air raids enlightened spectators as to both the presence of force the British aircraft could unleash, and the terror that enemy planes might similarly unload. The British air-control operations in Arabia (Satia 2008) brought a taste of the Orient to a London air-field in the form of a false Arab village ready for imminent destruction. In a strange simulation of the RAF terror raids, villages, pumping stations and other structures were attacked from the air.[19] A typically sanitized version of an attack upon an enemy village saw an enactment of British planes bombing the village only after its 'panic-stricken' citizens had heard the klaxon horn, escaping the 'overwhelming charge of explosive. The Village is practically demolished by the first attack.'[20]

Large formations of aircraft on fly-pasts were organized to copy the Russian and German masses of bombing aircraft (Fritzsche 1993; Palmer 1995), for it was supposed that '[p]eople who have seen them always say how impressive they are'.[21] This choreography of national force was even more cinematically portrayed (Palmer 1995) in the lighting effects displayed in an exercise of night flying in Hendon in 1921, during which the Aerodrome presented 'a perfect blaze of light'. The event saw Hendon opened at night with search-lights aimed at aircraft in the sky to guide them to the airfield. The aircraft themselves were outlined with hundreds of electric lights. The enclosures and the bandstand were even fitted with coloured lanterns, presenting the aero-drome with 'a remarkably novel and beautiful scene', lit up later in the dis-play by fire balloons and fireworks, illustrating 'War in the Air'.[22]

Airshows were not only popular in Britain. Intended to impress the Nazi ideology upon the visiting spectator's soul, the political and stage-managed events held at the Tempelhof airfield in Germany on 1 May 1933 did just that. The spectacle of flight became inextricably linked to the Nazis' propa-ganda campaign and the rejuvenation of a German national identity (Fritzsche 1992, 1993). David Pascoe (2001) uses the analogy of the cinema to com-pare the theatricality of the May Day rallies to the art of the movie theatre. These techniques would become commonplace in aerodrome and airport design as both displays and airports strove to build tension and drama almost as effectively as their cinematic counterparts.[23] Alternatively, the object of the air displays was to direct a visitor's focus not necessarily towards the aircraft, but rather towards the airfield and the activities of the personnel who worked there and maintained it. The airshows created a home not for the individual but for the collective community – for the *Volk*, for the nation. The airshows could envelop the spectator in the environment of the air.

In Britain the Empire Air Day displays were organized by the Air League, who sought to 'get the public *inside* aviation rather on the lines that the Navy say, "Come and see your Navy"'. The Air League implored the Air Force to open up the service aerodromes to the public for the yearly event, such that the public would 'have freedom of the place'. During these events visitors were to be shown round the sheds, see aeroplanes up close, look in the repair shops. 'The whole idea is not to keep the public behind fences but to let them roam about like they do on aboard the battleships.'[24]

The plan was essentially to give the public 'an insight into the everyday life of the Royal Air Force'.[25] The public could inspect the workshops, the school, barracks, playing fields, to see the various stages of technical train-ing and the way of life for the apprentices.[26] The intention was to open the airfields up so much that the League found it relatively difficult to secure against the risks involved.[27] J. M. Spaight, the Air Force lawyer, dealt with Lloyds of London on this very issue.[28] Adverts emphasized the ability for visitors to 'get behind the scenes of Aviation', to see 'how the RAF lives and works', 'how the great air liners enter and leave the airport with clock-work precision', and 'how clubs and schools train young men and women to master the air',[29] all publicized and distributed through a viral network of advertising, newspapers, cinema trailers and word of mouth.[30]

Over 50 RAF airfields were opened up to the public. Abingdon, Biggin Hill, Farnborough, Cranwell, Gosport, Filton, Henlow, Kenley, Larkhill, Tangmere, Waddington, all opened their doors, demonstrating 'photographic, armament, anti-gas, fire drill training'.[31] The airshow part of the day was to be limited to a few hours with specific items such as formation, drill, communication, message interception, supply dropping, dive bombing and fly-pasts.[32] The day effectively gave the public a different and more diffuse way in to aviation. Rather than centering the contact in the South East at Hendon, the events permitted a 'more intimate and more informal inspection of Royal Air Force activities'.[33]

The displays were necessarily spectacular. But as with Hendon the number of tricks and flourishes were toned down. It was argued that because the stations should be considered as 'at home', all flying activities should be as routine as possible. Letting the public see how the RAF 'lives, works and plays' meant that a sense of normality was needed. A memo advised that 'departures from reality should not be permitted to a degree which introduces an atmosphere of artificiality merely in order to attain sensational effects'. However, although the displays were themed on the idea of being 'at home' and opening up local RAF stations to their communities, their ambitions composed a far more extensive sphere of influence. Symbolized best by the souvenir programme produced by the *Air Review*, the airshows aimed to stimulate interest in British aviation and its capabilities as a strategic and global force.[34]

The politics of design and comfort

To impose an ideological message and generate a collective being-together at the displays, the interests behind the airshows required a public to consume them (Demetz 2002; Pascoe 2001). Through the design of and facilities around the Hendon airshow, its organizers tried hard to create this public – a spectating airshow-goer and a willing audience. Reports in the newspapers wrote up Hendon as 'London's most popular rendezvous', creating a vast social space for a 'pleasant afternoon's recreation', whilst fulfilling the 'desire at the same time to keep in close touch with the progress of aviation'.[35] Hendon could fill up free leisure time on weekend afternoons whilst providing a 'refreshing change from an ordinary outing' and supplying the 'interest and fascination that other London rendezvous lack'.[36] The dance in the sky was to mirror the ballet of sociality on the ground below.

As the aerodromes took on the role of a 'social centre', the publics they sought to grow and attract were stratified along the lines of a class politics that was expressed and reproduced in what Paul Virilio (2005) has identified as a 'politics of comfort', or what geographer David Bissell (2008) terms an 'aesthetic sensibility' and 'affective resonance' induced between bodies, objects, environments and other people. Magazines would convey the social scene evident at an airshow, finding no 'more fashionable crowd' than at the Hendon aerodrome. With 'a queen on either hand[, t]heir Majesties Queen Amélie and Queen Augusta – both of Portugal', one was surrounded by wealth, celebrity and the markers of sovereignty.[37] With the processual civilization of its airshow on the 'wilderness' of the landscape, Hendon was characterized as a 'resort of fashion – a veritable "Ascot" in London – a splendid health-giving – interesting – pleasurable – worry-not getting rendezvous'. Furthermore, 'everything that can possibly be thought of for the comfort of visitors' had either been 'done' or was 'doing. And the world of fashion in London has received it with open arms.'[38]

The show's organizers took many steps to study the comfort and convenience of visitors in 'almost every way', so engineering specific aesthetic and affective circulations between the airfield, things and other spectators through the management of touch, sensation and proximity (Bissell 2008).[39] The facilities were designed with an optimal position for the view of the aerial activities. The floor of the refreshment tents was deliberately raised several feet from the ground so that 'whilst tea and refreshments are being enjoyed, the flying can be observed with comfort'.[40] The show's organizers built a temporary display architecture of '[s]plendid and comfortable tea pavilions' designed with 'little striped red and white garden tents scattered about'. Each tent had its own 'wooden floor, and its dainty tea service prettily and invitingly set out'. The airfield provided plenty of walking space, 'where one can promenade, if so minded', but if not, there were 'thousands of comfortable chairs – arm and otherwise – for the weary, and plenty of fine flying". All this and more: 'Music, fashion, sport, interest, comfort, fresh air' were combined, 'what more can one want?' asked the Air Display Programme. 'Truly, on a sunny Sunday afternoon Hendon is a sight for the gods, and bids fair to be handed down to posterity side by side with Boulter's and Hyde Park.'[41]

The display organizers sought to entice spectators into the facilities to get as close to the aircraft as they could and in the utmost luxury. '[I]nfinitely more interesting than just watching an aeroplane as a speck in the sky', the aerodrome offered the ability, from any enclosure, to see pilots and passengers taking their seats in the aircraft, the mechanics going about their business and swinging the propellers, 'the swift rush over the ground, and then the graceful rise into the air'.[42] Seeing the aeroplane face to face, at a 'close view', could afford the 'visitor ample opportunities for comparing the various types of aeroplanes; also he can note how the rudders, elevators and "warp" [i.e. system for guiding lateral roll]' are controlled. How could one 'deny themselves the pleasure and fascination of watching flying and flyers at close quarters'?[43]

Figure 3.1 The meeting of car and spectator. (Reproduced with permission of the Alan Thelwell Collection.)

The airfields drew the crowds away from the city and into the outside air,[44] though some of the visiting public preferred to enjoy the sights from the perspective and comfort of their motorcars. Joined through the spectacular event of the airshow and the practice of the spectator, this collision of car and aeroplane cemented the modernity and inequality of flight (see Figures 3.1 and 3.2, from the opening airshow at Liverpool Airport, 1933).

Reports rendered an image of the meeting of mobile-machine-assemblage (Schivelbusch 1986;[45] Thrift 1996): '[T]here is no more comfortable vantage point from which to see the aerial evolutions than in a car,' an article in the *Court Journal* stated. This continued the long association between the corporeal stillness born of an act of sitting, the experience of 'comfort' and 'enjoyment' formed through the dance of aircraft above, the atmosphere of splendour, and the bodily work needed to *get* comfortable (Bissell 2008): 'To lean back among the cushions and watch [a pilot …] climbing to a height of 4,000 feet, and then making twenty-two consecutive "loops" in his descent', all this 'affords a series of thrills which can certainly be enjoyed nowhere else in such comfort'.[46]

Both car and aeroplane were objects to be consumed from without: the aircraft separated by their verticality and the cars by the horizontal distances created by 'specially arranged' facilities. There 'is no possibility of getting

Figure 3.2   Cars and spectators face the display. (Reproduced with permission of the Alan Thelwell Collection.)

wedged in among the crowd', a report mentioned. With the car forming a bubble between its occupants and the crowd, 'it is the easiest thing possible to get out at any moment when duty or pleasure necessitates a return to town', whilst tea and refreshments could be delivered to the car at 'a very moderate charge'.[47] If one wants to avoid the 'country roads', so 'crowded with holiday-makers' and traffic, 'there can be no better use to put a car to for an afternoon than to take it into the enclosure at Hendon'.[48]

The politics of comfort evident here had surfaced in the organization and planning of public spaces, particularly transportation infrastructures such as railway stations (Schivelbusch 1986) and other spaces of public voyeurism (on the theatre or cinema hall, see Bruno 2002). The public crowd was often conceived in contemptuous terms, as exemplified in Harry Crosby's recount of Charles Lindbergh's plane as it swooped down and across the field on its arrival in Paris in 1927:

> [A]nd there is pandemonium wild animals let loose and stampede towards the plane and C and I hanging on to each other running and the crowd behind stampeding like buffalo and a pushing and shoving and where is he where is

Lindbergh where is he and the extraordinary impression I had of hands thousands of hands weaving like maggots over the silver wings of the Spirit of Saint-Louis. (cited in Pascoe 2003: 24)

The regressed crowd's frenzied pollution of the purity of Lindbergh's mobility – flight's ignition of the scattering and stampeded collective – would be mirrored in the predictions of the falling bomb discussed later in the book.

Class and regional identity politics were stirred up in *The Aeroplane* magazine, publishing a story of one of the first flight meetings to be held in Britain at Dunstall Park, Wolverhampton, in 1910. The article questioned how wise it was to hold a meeting at the park with a crowd believed to hold questionable morals and discipline:

[A] crowd consisting very largely of iron and coal workers form the Black Country and a mob of factory hands from the potteries [would be] an even tougher lot to handle than the Lancashire mill hands at Blackpool, and as Dunstall is at the back door of the Black Country the very roughest of the rough are likely to be there.[49]

The report was prompted by an earlier meet at Worcester. Suffering from poor weather conditions, the airshow had closed midway, to the chagrin of the public. Invading the course, the crowd encouraged the aviators to appease their demands, leading to several members of the public being injured by dangerously flown aircraft.

Taking considerable offence at the proposition, the article received some attention by the local Midlands press. Yet an editorial on the matter did not simplistically rebuff the concerns expressed. The crowd, the article submitted, was a most difficult object to control; a mobile mass of unruly and susceptible characters, nothing, the editor wrote,

is more dangerous to the individual than the movement of a crowd of which he or she forms a unit. And so it is necessary to keep the crowd within bounds in order to prevent it from hurting those who are in it, but do not wish to cause any disturbance, and even to protect those who wish to create a disturbance but are too ignorant to foresee the consequences of doing so.[50]

Posing a rhetorical question to the city's Chief Constable, the article asked whether he would, in the event of a disturbance arising, prefer to 'deal with the more or less agricultural crowd of the Worcester district, or the mine and factory workers of the Potteries and Black Country?' On the other hand, the débâcle positioned *The Aeroplane* magazine's view as 'an illuminating contemporary attitude to manual workers: a view perhaps retained by some Southerners who seem to think that civilisation ends north of Watford'.[51]

Back at Hendon the spectator's education in aviation had to be directed. Information was displayed by event boards, bombs highlighted the start and end of the display, flags were used, black balls were raised on pylons, bells notified the start and end of each event. For visitors to become 'true connoisseurs of flying', instruction had to be given and treated as advice. Commentators had marvelled at the aerial grammar visitors were gaining. Among the 'cheaper enclosures', reporters talked of hearing 'really interesting discussions on the way the different pilots handle their machines'; these included:

> the effects of gusts at the different corners, according to the direction of the wind, and the way in which the pilots negotiate them, the way in which landings are made, the way a pilot executes a spiral, the way he takes his machine off the ground, and all the thousand and one little details of flying discussed as intelligently as a few years ago one heard the now vanishing crowds at cricket matches discussing the batting, bowling or fielding of the various players.[52]

Appraised for caressing the spectators with the intelligent details of the display whilst thrusting information down their necks 'in spite of themselves', the loudspeaker man was a key figure in any flying display. Visitors could even listen to instructions given and received from the aircraft's wireless transmission. Those who paid would be able to speak to the machines.[53]

The airshow was spatially, temporally and socially designed. From the chair to the airfield's architecture, it resembled the kinds of sensory apparatus described by Jonathan Crary (1990; Bruno 2002) which created material and affective affordances towards more optimal forms of spectatorship as well as proper forms of society (Pearman 2004: 38). It was a technique of aerial indoctrination on a grand and intimate scale (Crary 1999), 'inculcating specifiable forms of conduct and comportment' (Edensor 2001: 64). However, for some lucky spectators – as many as the famous aviator Alan Cobham could take up with him in his plane – the events of the airshow broke away from the sensibilities of stillness and comfort. 'If you have never had a Passenger Flight you have missed a 20th Century thrill,' airshow programmes enthused.[54] Passengers and pilots tried hard to convey the sense of flying (Langewiesche 1998), the feeling of ascent proving 'almost impossible to put into ordinary commonplace words'.[55] Spectators who were taken up recounted an array of sensations, emotions and affects. Leaving the ground, being unnoticeable and light, the coldness lost through an overpowering sense of excitement; an 'exhilaration which cannot be described in words; even when actually experiencing it the mind does not register more than confused impressions'.[56]

Some attempted to find anything comparable to the alien experience: 'The nearest approach to the sensation is that of battling with a rough sea on a windy and cold day on the east coast.' These descriptions were relayed by a Miss Jacqueline Roberts, an ordinary member of the public who won a free flight from the *Daily Express*. Looking down on the people, she found that they all looked 'so small and insignificant that one felt fearfully disappointed coming down to earth and prosaic matters again'.[57] Others were impressed with its departure from more terrestrial forms of travel such as the motorcar:

> The smooth and seemingly effortless speed of flight the absence of jolting or vibration such as of wheels passing along a road, which are the accompaniment of journeys by motor car; the magnificent panorama that is unrolled on the land below – slow passing, far distant, but revealed in marvellous detail; above all, the complex triumphant sensation of riding, surely and swiftly, through an unseen element – these combine to produce a joy so deep that it is dumb.[58]

For famous aviators such as Alan Cobham, this was the way to give the public an *aerial baptism*, to make Britain air-minded. Cobham (1978) was interested in stimulating enthusiasm in the general public, to allow people 'to look down and identify places on the ground, including their own homes [...] they don't want to see wings and wires and struts' (cited in Cruddas 2003: 155). But Cobham was just as committed to solidifying this interest into something permanent and infrastructural.

Building airspace

Airports coalesced as more permanent constellations of the state and regional authority, holding together the idea of national airspace, but doing so on the ground. They were the nodes around, over and through which nationalist sentiments or even more regional and localized identities were formed (Smith and Toulier 2001). After initially struggling to develop the required interest and enthusiasm to build a system or network of airport terminals, in Britain the idea of municipal aerodromes only took hold in public discourse during the late 1920s. This stood in contrast to the more progressive nations such as France and Germany, who had already built extensive airport networks by way of government subsidies and support, *airport systems* which had not gone unnoticed by the British. The reach of Dutch airline KLM also threatened to dent Britain's pride, and in more competetive terms (Dierikx 1991), airports were connected up into chains, a link and node network that would both defend as well as potentially strangle competitors' aspirations in the air. The reach of Germany's Lufthansa (Fritzsche 1992) would symbolize and perform the tying together of the nation by aerial movement, as would Italy's Alitalia (Caprotti 2008).

On Tuesday, 5 November 1929, 250 representatives from many of the local authorities of England met for a conference on the 'Necessity for Municipal Airports', organized by the Royal Aeronautical Society, the Royal

Aero Club and the Air League of the British Empire. The conference aimed to generate interest in the compelling imperatives for the necessity of airport construction. One of the ways they intended on doing this was to foster friendly rivalry and pride between the city's civic fathers, whilst expanding the debate to a question of national endeavour. The Mayor of Westminster opened the conference with an announcement typical of the aerodrome debate: 'The air age has arrived.' With the Mayor setting up the air as Britain's own sort of *manifest destiny* (Cronon et al. 1992), the aeroplane was described as inevitable. Teleologically worded, his speech saw aviation as Britain's natural progressor to the sea. It was the 'duty' of the present generations to realize that potential and progress. Municipal aerodromes were inextricably linked to 'the prosperity of the nation. [...] Trade by air has already assumed such proportions that it behoves every town to prepare for its reception as a matter of course.'[59] The President of the Royal Aeronautical Society spoke of the progress of European and other flying nations. France had 'adopted a bold and far-sighted policy, believing to some extent in the freedom of the air',[60] while Germany had carried 111,000 passengers by 1928 and aircraft in the United States had travelled some 7 million miles.

In an address on trade by air, Lord Herbert Scott, President of the London Chamber of Commerce, observed that 'when the Chambers of Commerce in that country [the United States] undertook to make the American business man "air-minded" they were, in fact, making the whole nation "air-minded," and this is what we want here in England – for the whole British nation to be "air-minded"'.[61] For Eric Geddes, Chairman of Imperial Airways, the issue was quite simply 'entirely a matter of airports'.[62] Without airports airlines could not develop the services they might. '[W]hat was the idea of these airports? It was impossible to fly until we had airports to land upon,' said Alan Cobham.[63] 'Airports and still more airports' became the slogan of the conference; the airport cause was becoming the national cause.

Airports were planned initially as spaces that would support the nation's progress, although they were to appear through a more bottom-up process of identity building. Evident in the circular sent out in 1928 by the Air Ministry to all major cities, the country's need for the provision of aerodromes at a local level was clear. The benefits were set out in terms of economic development and national rivalry with Germany.[64] Gateways to the nation maybe, they were also cast in terms of their performance as the front door to the city. As assertions of national interest, municipal airports were encouraged to function as builders of civic pride. *Building* an airport, moreover, was cast as a process of significant progress. An expression of modernity and a sign of forward thinking, the temporary and theatrical design and comfort of the airshow apparatus were formalized into the architecture of the new airport terminals, in Britain and on the continent.

Initial discussion for the first municipal airports took this capability as a primary theme for the new infrastructures. '[S]ightseers should be considered,' argued Major Mealing, comparing the German Tempelhof invention of 'a most delightful enclosure where one can sit and watch the flying, have tea, listen to music and have a most interesting day's outing'.[65] In Europe Templehof set the benchmark in the paradigm of not only airshow choreography but also airport design (Braun 1998), whereas in the United States airports developed architectural forms which featured tiered grandstands more befitting of a sports ground. From these spaces people could be instilled with political messages (Bednarek 2001). Airports were closely connected to the 'park and city beautiful' movement as the airport took on the properties of a recreational venue, as aviation historian and scholar Janet Daly Bednarek notes (2005: 355–6). More importantly, spectators could be convinced of the potential of flight, bought into its safety and, furthermore, sold on their potential for progress.

Influenced by this progress, UK airport architecture proceeded through the work of the Royal Institute of British Architects' Aerodrome Committee and the Aerodrome Advisory Committee. Architect Graham Dawbarn and his partner Nigel Norman (Bingham 2004), who had visited the States in 1932 on behalf of the RIBA committee, saw great advantage in constructing terraces and observation decks. John Dower, Chairman and Secretary of the Aerodrome Advisory Committee in the 1930s, stated that the function of an airport as a recreational venue brought its own financial benefits to the local authorities. Dawbarn's American report would add:

> Roosevelt Field has produced an interesting system of 'bleachers' (or stands for spectators) and car pens involving a minimum of supervision, three of four thousand people being able to see the whole Aerodrome from their cars, the whole falling within the 'dead' area (aeronautically speaking) formed by adjoining buildings. At Cleveland there is a large grand-stand reminisce of Epsom, on the side of the field furthest from the other buildings while, though I did not see it, I understand that at Glenview, Chicago there is a pukka grandstand on top of an enormous hangar. (Dawbarn 1932)

As if at a 'drive-in movie' (Gordon 2004: 59), the airports Dawbarn and Norman created featured spectator facilities which were second to none.[66] At Birmingham Airport in the Midlands, the 'wide sweep of balconies resembling the promenade decks of an ocean liner' was designed in order to 'give a splendid view of the Airport and surrounding countryside'. Furthermore, 'a great cantilever concrete canopy on either side' of the airport was installed to provide 'protection from the weather to passengers and aircraft'. Under these canopies the largest airliners could take shelter.[67]

As we will see in the workings of the border, the airport flexes its nodes in order to establish a boundary. Manufacturing both real and imagined fields of outsides and insides (Feldman 2007), the outline and shape of the nation are projected from the airport and figurally from the airshow. Whilst this premise is well established, in the following section I argue that such a process of bordering is radically uncertain and almost continuously undone. For Virilio (2005: 98), the airport's role as little 'more than a projector' has always offered up this tension – an *airport politics* of the anti-city which 'opens on the nothingness of a territory disappeared' (2005: 103).

Undoing Aerial Space: Post-nationalism and Projective Power

The air display and the airport were the nuclei around which national identity could condense and coalesce. They formed a territory for belonging. In turn, early socio-legal debates over geographical air-space expressed similar scalar relations of national sovereignty (Blomley 1994). In these debates, French jurist Paul Fauchille (1901) proved to be particularly influential. Because an aircraft (Fauchille was writing about airships) could not physically occupy the sky, neither could it possess it. These arguments depended on direct comparison of the air to the materiality of the ocean:

> By its immensity and its fluidity the atmospheric layer resists all possession. It is, materially and physically, impossible for a people, even with all the forces of the world at its disposition, to exercise over the air an effective grasp, to mark it with the seal of its authority. (Fauchille in Banner 2008: 49)

'Necessarily ephemeral', the sky was not substantial enough to be maintained as sovereign territory.

For others, the possibility of effective occupation and therefore possession was entirely feasible. Arthur Kuhn, an American lawyer specializing in international law, suggested that once aviation had become commonplace, 'states will be able to execute their will upon the zone abutting them from above' (cited in Banner 2008: 52–3). From the principle *cujus est sulum* it followed that a sovereign could allocate rights in the air. From property law, the maxim presupposed territory with three-dimensional or volumetric properties. Like a lighthouse beam cast upwards into space and down to the core of the earth, property and territory were projections which extruded out in three dimensions. Airspace was a volume without visible limits or definition, and therefore from the 'conception of boundaries' as a 'lateral description', it was difficult to grasp 'mentally'.[68]

Both points of view met head on in the aerial navigation conference held in Paris in 1910. Having been asked to respond to a questionnaire, the British, who were initially eager to take a back seat and listen, found themselves 'impelled to take a more active part in the proceedings'. The 'extreme liberality of the views of some foreign countries on the question of the admission of foreign air-ships', they wrote in an interim summary, 'seemed to demand very careful consideration, as it closely affected important national interests'.[69] The zeppelin menace overshadowed the British position, forming what David Butler (2001) calls a 'technogeopolitical threat'. Aircraft or aerostats as they were discussed at the conference threatened the security of the state: '[T]here can be no user of the air which does not to some extent affect the security or the administration of that State.' Anticipation of the dreaded aircraft encroaching on vulnerable populations on the ground below served to ensure the British argument was one of shoring up the space above their territory, and thus a natural continuation of existing international law (Butler 2001: 641). Alternatively, the French and German governments had much more to gain by opening up airspace to the advantages they held in aircraft technology. The British fear over their enemy's technologies acted as accelerant to their own development of aerial progress.

Air law held together a relatively obdurate space by attempting to grasp not at the gaseous materiality of the atmosphere or the mobilities of projectiles, but at the social relations between these subjects and objects. As Michael Milde writes, 'law does not govern "objects" as such' (2008: 1), but only social relations. In just this way the national space of the airshow and the airport was imprinted and materialized in vertical territory. Understanding the relation between aerial flight and nationhood as particularly antagonistic, this section of the chapter explores the undoing of aerial space in two ways: firstly, into a putative and post-national terrain of flexible citizenship and free mobility; and secondly, into a disintegrated patchwork quilt of uneven authorities and overlapping sovereignties.

Open skies

Although the influence of Hugo Grotius' *Mare Librium* (The free seas) in international relations never found its form in the original production of airspace architecture (Bull et al. 1992), moves are afoot radically to alter the structure of the sky, opening up borders and territories to governance at an extra- or post-national scale, whilst removing national and historical restrictions through deregulation, liberalization and bilateral agreements. Take rhetoric that has underpinned the formation of a European space of unrestricted mobility between member states into what Ole Jensen and Tim Richardson (2004) call the European ideal of a *monotopia*. In the emerging discourse over a Single European Sky, the shifting ideological underpinnings of air-travel mobilities from the post-war period have moved away from the bounded fortress formulations which mimicked the territorial borders of the nation-state (Pascoe 2001, 2003; see also Kaplan 2006a), to something closer to the Grotian ideology of the seas. As Jensen and Richardson argue, what is on the agenda is the 'mending of the previous "partitioning" of Europe's skies', articulating Europe as 'one space, at multiple scales – a space of monotopia' (2004: 82). Following the formation of a single European market that 'eliminated borders on the ground' and 'dismantled economic borders', European stakeholders have turned towards the United States in order to make the argument 'that borders in the sky should not exist'.[70]

The international organization and governance of aviation is clearly not anything new. The formation of the International Civil Aviation Organization (ICAO) in 1944 sought to maintain national sovereignty and responsibility for airspace and aviation security – even health security (Budd et al. 2010 in press) – yet it also imposed the consensus of internationally agreed obligations and the identification of 'regions' with common conditions and contexts. In the United States, the model of a national management of airspace has evolved from a National Airspace System into the Next Generation Air Transportation System, which has sought to do away with the regional- or state-level management of airspace.

In Europe, agreements began in the 1960s as representatives of Belgium, France, the Federal Republic of Germany, Luxembourg, the Netherlands and the United Kingdom met in Brussels to sign the EUROCONTROL International Convention relating to Cooperation for the Safety of Air Navigation. The idea of EUROCONTROL was to create an upper layer of airspace across Europe that was truly international and controlled by an international system of air traffic management. Sovereignty of these spaces was held until February 1972, when the Maastricht Upper Area Control Centre entered operational service. The UAC covered the upper airspace of Belgium, Luxembourg, the Netherlands and northern Germany and was organized by a truly international air traffic control which marked, for the first time, how air traffic in one country could be controlled from another.

Pressed on the grounds of efficiency and cost savings, and following the path of the American NextGen, the proposals for Single European Sky and Single European Sky II have stayed true to the original principles, but shifted emphasis towards the environmental benefits a Single European Sky would provide in terms of carbon emission and fuel savings on individual flights. This is quoted at around 10 per cent. Air traffic management costs could be cut in half, and the capacity of the skies, it is argued, would be increased by about 300 per cent. The IATA's Director General, Giovanni Bisignani, has urged governments 'to put aside local politics and focus on what is best for Europe and for the environment'.[71]

The geographical imaginations involved in Single European Sky are not limited to a rendering of a metaphorical 'single sky' overcoming all borders and obstacles, but incorporate a much more flexible conception of airspace which is not static, but reflexively responsive to the desired vectors of aircraft. This is being founded on the back of new air traffic management concepts known as 'trajectory-based operations'. Described as a significant paradigm shift from an 'airspace environment' to a 'trajectory-based environment', a flight within the Single European Sky will fly not according to partitioned airspaces along lines of sovereign territory, but it will collaboratively set its flight trajectory, achieving preferred routing with 'no preferred routes'.[72]

The shift to these new spatial or vectoral aerial architectures which are eradicating formalized container-like definitions of territory is occurring hand-in-hand with movements on the ground. Europe promises a new and putative 'free market' of aerial movement across the architecture of the sky and national regimes of mobility citizenship (Brooker 2003). Single European Sky follows the partial creation of a free-European area of mobility citizenship known as Schengen. First begun in 1985, and eventually formalized into the EU and amended by the Lisbon Treaty, Schengen offers a way for the citizens of member states to travel uninterrupted across borders. Under article 61: 'The Union shall constitute an area of freedom, security and justice with respect for fundamental rights and the different legal systems and traditions of the Member States.'[73] Article 62(a) meanwhile, ensures 'the absence of any controls on persons, whatever their nationality, when crossing internal borders'. Playing a fundamental role in the facilitation of Schengen passengers, European airports have been fundamentally altered in the way they deal with passenger flows, immigration checks and other biopolitical forms of securing. Quite new airport architectures have emerged allowing travellers to traverse within post-national Schengen space separated from other non-Schengen passengers, existing within a space devoid of national markers that would have previously impeded their transit.

Although Schengen appears to open up national boundaries to a European community of mobility, it actually serves to create new fluid borders around the extra-territorial state of Europe (Delanty 2006). Like airspace, Schengen aims to maintain the 'fortress' boundaries of its member states, whilst actively managing trajectories or vectors of movement that extend lines of securitization within and beyond the Schengen area. Étienne Balibar's critique of Schengen's 'European racism' (Balibar 2004) highlights these dual policies. Schengen works by both ring-fencing its area from outside travel, whilst it intensively differentiates the mobilities and immobilities of certain travellers within and without its borders (Walters 2006). By the mediation of the state and Schengen's supra-national practices of movement, Balibar's 'fundamentally unconscious' character of foreigner or stranger is constituted. The mundane movements of Schengen space are ' "lived" distortedly and "projected" as a relationship to the Other' (Balibar 1991: 15).

As Gallya Lahav and Virginie Guiraudon explain this process in more detail, controlling the movement of EU peoples increasingly takes place away from the border and before 'undesirables' reach EU member states. A wide range of airport, airline and civil personnel are 'urged to reach deep into societies to uncover undocumented foreigners, deter asylum-seekers and prevent the exit of the "huddled masses" ' (Guiraudon cited in Walters

2006: 194). Protecting the integrity of the EU therefore involves penetrating within societies both inside and outside the Schengen area (Broeders 2007). The result is a form of management that is based, like the sky above it, on trajectories and vectors of threat. As the Schengen Information System (SIS) moves from its first-generation system of 'reporting' to one of reporting and 'investigation' in SIS II, border agencies and national police forces will be able to perform checks on a plethora of information, including alerts for firearms and criminal records, missing and wanted persons, stolen vehicles and witnesses. This includes the UK, which is not a Schengen member. The European Commission's agreement with the United States has created a similar transatlantic architecture of trajectories of information flow criss-crossing the ocean before the passengers do. The arrangement sees data 'pulled' from European carriers 'pre-emptively' (Amoore and de Goede 2008),[74] extending the function of the airline check-in desk as the initial gateway of territorial access (Gilboy 1997). What Kenneth McDonagh (2008) characterizes as an 'exercise in transnationalization' has clearly surfaced in this coordination of resources; even within Europe, members of the European Civil Aviation Conference (ECAC) security audit programme have begun harmonizing their national security practices.

The result of all of this is the neoliberal emphasis on enhanced mobility and flexibility across space. The problem is that is incredibly uneven. Peter Andreas' (2003) term 'rebordering' is apt especially within Europe given how the transnationalization of its borders enhances the mobilities of its member citizens while prohibiting movement of those not classed as such – a trend that Matt Sparke has described as a process of 'redrawing' 'lines that shut out and imprison diverse others' (2006: 176) – creating a status of insider and outsider (Agamben 1998: Bigo 2006). Taking the trajectories of movement helps reveals an uneven geometry of territorialization and sovereignty bent and stretched through the movement of borders and authority.

Uneven skies

The state's affinity for 'remote control' (Guiraudon and Lahav 2000) is particularly evident in airport pre-clearance programmes. As Mark Salter (2004) proposes, our borders are actually decentred or 'delocalized'. Tested in more recent forms of airport borders, Nick Vaughan-Williams (2010) describes new 'alternative border imaginaries, vocabularies, and conceptualizations'. Focusing on Britain's policy towards borders after the London bombings in July 2005 and the 2007 attack on Glasgow Airport (Leppard 2007), we see new ideas about the airport/border being proposed through an enhanced language of border security which has less and less to do with container-like territorial control and more to do with bio-political management.

As Vaughan-Williams (2010) explains, the UK Home Office have argued for a wholesale change in how we think of the border, away from 'a single, staffed physical frontier, where travellers show paper-based identity documents to pass through', to one that will respond to the 'the step change in mobility that globalization has brought'. In other words, '*a new doctrine is demanded*' (Home Office 2007: 3, emphasis added in Vaughan-Williams). Now the territorial border is no longer 'a fixed line on a map', but rather a 'new offshore line of defence' (Home Office 2007: 2).

The emphasis of all these reimaginings is to distance or dislocate the border from territory, to push the identification and securitization of individuals to an arm's length. 'Exporting' (Rumford 2006) the aerial border means risky individuals can be checked at a distance rather than when they reach the UK's shores, at which point it is 'too late'. Such a process is constituted by airport security officials stationed in other airport borders, or decisions over one's ability to travel taken away before one's flight. These kinds of dislocation are coming to form part of a wider logic of pre-emption as the airport and the aeroplane are entangled in differential vectors and spaces of decision making and information (Amoore and de Goede 2008).

We will explore just how these decisions are being made or indeed *deferred* later on in the book, but it is worth discussing that the off-shore airport border is neither new (at least since July 2005), nor is it predicated on the sophisticated 'technological' fixes of information sharing and biometric technologies. Racial and ethnic discrimination was levelled at the British government's border agencies in a 2002 claim to the British High Court, due to the agreement that gave British immigration officers permission to undertake immigration scrutiny at Prague Airport in the Czech Republic, in order to 'pre-clear' passengers before they boarded flights for the UK (*Regina* v. *Immigration Officer at Prague Airport and Another*).[75] Czech Roma passengers, 'who [had] not left the Czech Republic nor presented themselves, save in a highly metaphorical sense, at the frontier of the United Kingdom', were denied express permission to visit the UK or were treated unfavourably.[76] The evidence of the European Roma Rights Center indicated that a Roma passenger was 400 times more likely to be refused pre-entry clearance to the UK than an individual non-Roma. The Roma were subjected to longer questioning by immigration officials than the non-Roma, and 80 per cent of Roma were subjected to a second interview as compared with less than 1 per cent of non-Roma.

Although the appellants were successful in their charge of racial discrimination, their first appeal failed. This first charge argued that the UK had a domestic and international obligation to the principle of non-refoulement – in other words, that someone who has left their first state on the grounds of persecution should not be rejected and returned without appropriate enquiry into their fear of persecution (Kesby 2006). Their argument rested on whether the Roma passengers had actually left the territory of the Czech

Republic. The implication of this position was: had the location of the UK borders moved to Prague Airport, albeit in a very ad hoc way, given the temporary operation of the UK pre-clearance programme? And if so, did the UK and international obligations to non-refoulment apply extra-territorially? What is so crucial and interesting for our purposes in the context of understanding borders in relation to air travel is what the court contest meant for the way the border is imagined. Although their claim was rejected, the court had to decide where the Roma passengers were placed in relation to their own country's borders and those of the United Kingdom. Whether they were located inside or outside a particular territory made a dramatic difference to their ability to claim asylum or international human rights. Whilst Lord Bingham drew upon a difference between an actual and metaphorical border, legal scholar Alison Kesby suggests that 'it is arguable that the Roma were confronted with the "UK border" while still in Prague. There was a disjuncture between the location of the "geographical" border and the place at which the border was experienced. Immigration rules which otherwise would have operated at UK ports were to apply extra-territorially' (2006: 8).

The Roma case did not set a precedent in international aerial borders either. Since 1952, the international airport system has permitted the extension of sovereignty through pre-clearance programmes in Canada. Formalized in the 1974 Canada–United States Air Transport Preclearance Agreement, travellers to the United States may be 'cleared by American authorities while still inside the Canadian terminal'. In this sense, the borders of the United States do not remain at its geographical limits, but are extended over 'variegated authority and overlapping sovereignties'. Performing these complexities through sets of procedures, processes and blurred authorities, Salter writes,

> [i]nspecting officials may frisk, but only host officials may strip; inspecting authorities may detain (secondary inspection, though mandatory at the discretion of the officer, is not legally detention), but only host authorities may arrest; inspecting officials may examine aircraft outside of the preclearance area, but only host authorities may escort a passenger back to the preclearance area for further (re)examination. (2006: 56)

Whilst border agencies work to maintain the power and sovereignty of the Canadian state to set and define the laws of the national territory, Salter argues how such a rendering is simply 'shattered by a purpose-built American border post' within Ottawa's airport and others such as Calgary, Edmonton, Vancouver and Montreal. In what Salter describes as the 'up-streaming of international borders' is created a 'concrete example of the deterritorialization of sovereignty, where a state enjoys authority and legal precedence outside of its national territory' (2006: 56).

Between Canada and the United States, pre-clearance is increasingly unbundled and rendered uneven through other sorts of agreements designed in order to facilitate 'trusted' and 'low-risk' mobility. NEXUS, for example, elicits movement between a regional zone of space that crosses both the United States and Canada, a tunnel space (Graham and Marvin 2001) that burrows through sovereign territory. Movement within Cascadia, the cross-border region that stretches in an affluent imaginary geography between Vancouver and Portland, Oregon, is made frictionless by expedited security lines and enhanced immigration queues. Members pre-cleared by submitting their details in order to join up to the scheme effectively purchase their post-national citizenship (Ong 1999, 2006), which affords them their faster mobility through dual belonging.

If the tendency has been to open up aerial space incredibly unevenly, or, to be more precise, through an essential extension or off-shoring of the border under the umbrella of securitization, airspace is being unbundled in just the same way. Airspace is unevenly spaced with different states extruding into one another. This takes place across a number of dimensions and in a manner that reconfirms our understanding of airspace as volumetric (Graham 2004). The temporal and spatial extension of the airport/border cannot perfectly map onto the geometries of airspace. Rather, under the guise of security and freedom in the 'war on terror', aerial territorial integrity is displaced as sovereignty has been made contingent (Elden 2005, 2007). Common conceptions of territory are under tension (Elden and Williams 2009), for as a contingent entity airspace is violated by odd and sometimes unlawful unilateral military infringements into the sovereignty of one nation's airspace by another (Williams 2009).

Alison Williams (2009) has argued that no-fly zones are a clear example of this tension, although their legality has been rather muddy. Even if their purpose has been initially humanitarian or for security – as understood outwith the sovereign nation – no-fly zones have often become platforms for violence and aggression, dismembering territorial integrity (Bernard 2004). Under the charge of providing humanitarian support to the Iraqi population, the first Gulf War of 1991 saw the US and the UK tear strips into Iraqi airspace in both Northern and Southern no-fly zones. Unsupported by a UN Security Council resolution or international law, the zones have come under criticism for their violation of Iraqi sovereignty. As an 'occupation', a no-fly zone restricts the sovereign's air power from occupying their airspace and, thus, controlling their territory. By denying a nation's aerial sovereignty, a no-fly zone can have dramatic effects upon an adversary's 'political, economic, military and human conditions' (Jouas 1998).

Iraq's no-fly zones relied upon a murky logic of international law under the United Nations Charter, and have been used for both humanitarian purposes and the delivery of aerial bombardment (Grosscup 2006). Attempting to protect the 2 million civilian Kurd population who were escaping Iraq, and fleeing helicopters and fighters deploying gunfire, napalm and phosphorous bombs, the US relied on the provisions of UN Security Council Resolution (UNSCR) 678. Passed to restore peace and security in the area to secure an airspace of all fixed rotary wing aircraft north of the 36th parallel, the conditions of the no-fly zone were relative. Operation Provide Comfort saw US aircraft deployed in the no-fly zone in order to provide air drops to the fleeing Kurds. Recent Bosnian–Serb conflicts have required the same manner of protection through the modification of aerial space orchestrated in a digital virtual war (Der Derian 2001; Ignatieff 2001).

After the disengagement of ground troops, the no-fly zones allowed US, UK and French forces to top-slice Iraqi sovereign territory. Holding the high ground would mean enhanced control of that below. Enlarged no-fly zones in 1994, intended to reduce Sadam Hussein's ability 'to threaten his neighbours and America's interests', over-cut Iraqi airfields that now fell underneath the expanded zones. Although officially Iraq's sovereignty was meant to remain intact, what Williams (2007) calls a 'vertical techno-geopolitical practice' did quite the opposite. And in so doing, the practice prevented Iraq from 'effectively exercising the territorial preservation of its existing boundaries, or adequately exerting its control and authority over the whole of its territory' (Williams 2007: 521). Indeed, the apparent failure of the no-fly zones to halt the humanitarian crises mentioned has added further fuel to those who protest the necessity of their use in Darfur (Zenko 2009).

As with the border, the lines of aerial space require maintenance and enforcement. The UN and US relief operations to Sarajevo in 1992 meant a huge-scale humanitarian response mission which required European and US daytime and nighttime air-lift and air-drop operations which were subjected to fire from Serb aircraft. UNSCR 781 effectively banned all flight in Bosnia by military aircraft that were not assigned to UNPROFOR. But the ban was ignored. Serb aircraft took little notice of the zones. Thus, the declaration was later enhanced by the UNSCR 816, which saw NATO enforcement of the no-fly zone by Operation Deny Flight. No-fly zones therefore require constant force in order to sustain their existence. As Williams argues, mobility is a fundamental trope to be considered in the projection of the aerial border. For while the border is apparently 'static, unmoving, fixed', it is also preserved by aircraft 'being in continual motion [...] an aircraft ceases to be a useful tool of power projection if it is not in flight' (Williams 2007: 509).

No-fly zones require further cooperation between national and international agencies, as shown in the 1999 Kosovo conflict, during which the space above Bosnia and Herzegovina, Croatia, Hungary, Italy, Romania, Bulgaria, Slovenia and the former Yugoslav Republic of Macedonia was closed. Not only did civilian aircraft require rerouting, but military and humanitarian aircraft requiring access to the airspace needed assistance through the civilian air traffic control system, both in and out of the closed airspace. NATO and the EUROCONTROL Central Flow Management Unit (CFMU) worked hand in hand to coordinate mobility within the conflict zone.[77]

Eyes from above

Much like the exported border, no-fly zones enable a distanced platform under which sovereign territory cannot be shielded or sealed up. Other state and territorial formations give legal rights for the surveillance and securitization of space that may well lie beyond the country's land borders (see Campbell on tele-vision, 2007, 2009). Gaza is a case in point where domination is enforced not through asymmetric no-fly zones but rather through the legal sovereignty of its airspace by another state. Enshrined in the Gaza–Jericho Agreement of 1994 is Israel's continued oversight of Palestinian skies, a strategy it illegally repeats through its overflights of Lebanon's sovereign airspace:

> All aviation activity or usage of the airspace by any aerial vehicle in the Gaza Strip and the Jericho Area shall require prior approval of Israel. It shall be subject to Israeli air traffic control including, inter alia, monitoring and regulation of air routes as well as relevant regulations and requirements. [...] Aviation activity by Israel will continue to be operated above the Gaza Strip and the Jericho Area.[78]

With these security arrangements made permanent in the Camp David negotiations, by way of aerial surveillance and power, Israel commands its continual superiority of life on the ground from the air. In this sense, both Lebanese and Palestinian sovereignty are violated not only at a vertical layer, for the layer of airspace becomes 'primarily a place to "see" from', offering up an 'an observational vantage point for policing airwaves alive with electromagnetic signals – from the visible to the radio and radar frequencies of the electromagnetic spectrum' (Weizman 2002). Private and public spheres are penetrated from above through a 'vacuum-cleaner' approach' to everyday life:

> [S]ensors aboard unmanned air vehicles (UAVs), aerial reconnaissance jets, early warning Hawkeye planes, and even an Earth-Observation Image Satellite, snatch most signals out of the air. Every floor in every house, every car, every telephone call or radio transmission, even the smallest event that occurs on the terrain, can thus be monitored, policed or destroyed from the air. (Weizman 2002)

Netanyahu's remarks in June 2009 indicate the probable future for Palestinian sovereignty over this portion of their territory: 'The Palestinian territory', he stated, 'will be without arms, will not control airspace.'[79]

The idea of securing one's territory and land borders by policing another's airspace is even witnessed in schemes that align the priorities of the environment and national security together. *Sistema de Vigilância da*

Amazônia, or Amazon Surveillance System (known as SIVAM), as it was first named in the mid-1990s (Raytheon 2000), was constructed with the purpose of creating a surveillance net cast over the 2 million square miles of the Brazilian legal area of the Amazon Rainforest (de Costa 2001). It was touted as Brazil's 'eye on the Amazon'. Intended to monitor forest fires, deforestation and illegal logging along with drug trafficking, a system of aerial remote sensing has been constructed allowing the state 'more presence'. CENSIPAM, as the system is now known, employs Embraer aircraft equipped with various kinds of radar and sensors that hoover up different levels of the microwave, visible and infrared spectrum. The system penetrates the jungle canopy, cloud cover, fog, smoke and the weather conditions that cloak it, identifying 'targets that "cannot be seen"'.[80] The aircrafts' movements have helped to shore up the porous borders under Brazil's protection. Reports tell of drug smugglers using the Amazon's tributaries to traffic kilogrammes of cocaine, subsequently busted by Brazilian police, who are guided in real time by remote-sensing aircraft pinpointing a clandestine airstrip. A network of aircraft is connected with the CENSIPAM control centre, who coordinate their information with other governmental databases, satellite imagery and 'on the ground' police forces (Collitt 2008).

The security of the Amazon and the effectiveness of the SIVAM/CENSIPAM system are increasingly dependent upon international support, and not only in terms of financial assistance. An agreement signed between Brazil, Columbia and Peru in March 2009 has seen the declaration of proposals which will mean CENSIPAM is significantly extended to the surveillance of Columbian airspace, 50 km within its borders. Building on established Memorandums of Understanding between Brazil and Peru, Brazil's expertise in using and interpreting remotely sensed data will be shared with the Colombian and Peruvian governments, allowing each state to surveill the other's airspace.

Conclusion

The dance and verve of a plane in the sky appears so singular, individual and unbounded. But we know that this is far from the case. An aircraft moves within a domain of international relations, tight regulation, and it flies between strictly controlled invisible lines of borders and air-traffic control delimitations. Indeed, the movements that go on along the ground both support this mobility and mirror its submission to the sanctity of the nation-state. Engineered minor bodily affective states of comfort even express this submission and project the analogue form of national belonging.

On the other hand, the aeroplane reinforces these boundaries and definitions as it simultaneously probes, tests, resists and undoes them. The reorganization of airspace, new forms of political community, the over-cutting of territory, have all worked against the volumes projected by the individual nation. The trajectories of movement mediated by the aeroplane have shaped and reshaped the projected dimensions of national, political and cultural forms. As we saw with the scareships, the *affect* of the aeroplane has been the projection of feelings into the air, often a reflection of the aerial subject's concerns and fears.

Part Two

Governing Aerial Life

Chapter Four

Aerial Views: Bodies, Borders and Biopolitics

[I]t is desirable to draw attention to what may be termed the Air View. All who have flown must have realized some of the wonders of the panorama which stretches out before them. Yet few can have perceived all that the air view really implies.

(Bourne 1930: 17)

[As] organisms are examined under a microscope; the interpreters of these photographs work exactly like the bacteriologists. They seek on the vulnerable body [...] traces of the virus which devours her [...] which, under the lens, appear like tiny bacilli.

(Saint-Exupéry in Baddington-Smith 1958: 246)

To target something is to express a desire to possess it in some way.

(Fox 2009: 33)

Introduction

The aerial view is often taken to be epistemological (Kaplan 2006a). Donna Haraway's 'god trick' describes the imagined projection of a gaze from high above. In this violence of seeing from nowhere, vision 'becomes unregulated gluttony; all seems not just mythically about the god trick of seeing everything from nowhere, but to have put the myth into ordinary practice. And like the god trick, this eye fucks the world to make techno-monsters' (Haraway 1991: 581).

The aerial platform's eye has certainly fucked the world, as detailed by many writers and historians describing the violence wrought by the aeroplane in Dresden (Freidrich 2006) and Hamburg (Lowe 2008), London (Calder 1991), Vietnam (Gibson 2000), Iraq, Darfur and Gaza (Grosscup 2006). This is not the view of Haraway's (1991) deity, who is both everywhere and nowhere. The aerial view is in fact positioned, often taken to be from above looking down. Moreover, the gaze is filtered, distinguishing what 'is lived' and what is 'perceived' from what can be 'conceived' (de Certeau 1984: 38). For Henri Lefebvre (1991) and Michel de Certeau (1984), it is the 'lived' world of experiences, perceptions and sensations that is obliterated by the view from above – the view from the planner looking down from his skyscraper office – obliterates. Writing from the perspective of the World Trade Center in New York, de Certeau's elevation turns the city 'into a text that lies before one's eyes. It allows one to read it' (1984: 92).

It is not in the least bit surprising that a dominant use for the aeroplane has been a mode of knowledge capture. For James C. Scott (1998), the view from the aeroplane converges with the synoptic perspective of the statist planner, wishing to know all and to transform all in his masterplans. The aeroplane reveals in a way that makes it equivalent to the camera, the telescope and the microscope (Vidler 2002: 37). Take the aeroplane's perspective as Le Corbusier's muse and inspiration for his modernist plans of Brasilia. The architect, he wrote, 'was now endowed with a new eye: the eye of a bird transplanted into the head of a man; a new way of looking: the aerial view' (Le Corbusier in Vidler 2002: 39). From the aircraft the sort of 'rational intelligence' that would ordinarily be acquired via the collection of facts, analysis, education and comparison 'suddenly becomes a matter of total and first-hand experience for the eye' (2002: 39; Morshed 2002; Pinder 2004). Now, and from above, the eye is neither clouded with subjectivity nor distracted by feelings; it is calculated, it is truthful.

In this chapter we are concerned with how the way we see *aerially* has worked to produce aerial subjects and spaces mediated under a field of control, combining both the aerial survey and the airport biometric as an aerial gaze. Both, we will see, address space, populations and mobility together through contradictory and overlapping forms of sovereign and biopolitical power. The chapter will explore several techniques or 'ways of seeing', resembling what Geographer Nick Blomley identifies in the visual abstract space of the cadastral survey, which, through its various techniques and processes, 'helps make a world' 'that exists' (2003: 17).

Seeing the Wood for the Trees: Targeting, Administering and Managing Populations

On 10 November 1919 John Moore-Brabazon posed a question to the British Parliament in the House of Commons: 'What steps, if any, have been taken to utilise the experience obtained during the war in connection with aerial photography for mapping purposes?'[1] Surveying was becoming seen as the 'foundation of progress' in the colonies and in the conduct of modern and late modern warfare. But in the process of the aerial survey's enrolment as the 'weapon' of colonial progress, and, as we will see, in the utilization of biometric imagery for managing mobile populations, both their *fidelity* to the ground and their *resolution* of accuracy – their mode of what Lisa Parks (2002) calls tele-vision – came under question.

Aerial survey was born in the crucible of desert warfare during the First World War. It first emerged within a variety of different practical domains, enabling the construction of new detailed maps that facilitated the drawing of territory in military planning (Collier 1994, 2002). In Britain it was seen as 'one of the most urgent needs of today'. With 'special functions', how the 'new weapon' (Bourne, 1930: 15) could be exploited was a commonly posed question. Today, this kind of aerial knowledge is known as 'total information awareness' (Graham 2003, 2008): unmanned aerial drones, photographic reconnaissance (Campbell 2007, 2009), satellite telemetry (Parks 2002) and real-time information all enable the military to *immerse* themselves in the battlefield (Virilio 2005). The 'modern aeroplane is really a travelling observation platform moving at from fifty to 100 miles an hour', first summarized the Army's Lieutenant-Colonel Macleod (1919), who noticed the observant possibilities of the aeroplane's position. The emergence of techniques and technologies of seeing from the air moved hand in hand with the creeping movement of imperial exploration, colonial administration and development. Aerial survey and photography opened up the world to a distanciated gaze by telescoping these distances for the imposition and projection of power and reach. Furthermore, it focused, targeting its lens upon enemies and opportunities.

The view from above began to be imagined as a way to cover and penetrate the vast and relatively unknown geographic detail within East Africa, India, Egypt, Palestine and Iraq, as well as the colonies of East Asia. With this new brand of what geographical historian J. K. Wright (1952) called the 'substance of geographical knowledge', both the Royal Geographical Society and the American Geographical Society became increasingly interested in the possibilities of the aeroplane. With it, the British Empire could be tied together as new flight routes were discussed (Walmsley 1920), and new perspectives on physical geography could be seen (Lee 1920). The view from the aeroplane could bring a different perspective on the study of rural societies (Blache 1921; Tarde 1919), whilst new techniques in mapping were produced and demanded by the aeromobile navigator (Whitaker 1917; Woodhouse 1917).

The infrastructure of air-routes and pathways necessary to conduct aerial photography would have to be built, constructing a symbolic and material presence in colonial regions – the aerial survey as harbinger of *development*.[2] Requiring considerable infrastructure to support the mobilities and maintenance of aircraft, air survey was a conduit through which development could be piped.[3] The infrastructure necessary for air-routes and commercial services could already be in place for colony development. The aerial view revealed a reality that demanded improvement and development. As Christine Boyer

(2003) explains, for Le Corbusier in his *Aircraft* (1935) and *The Four Routes* (1947), the aeroplane 'enlightened us', archiving a documentary record of the miseries of urban existence which justified its renewal as the truth of the city could not be ignored. The aircraft 'observes, works quickly, sees quickly, never tires' (Le Corbusier in Boyer 2003: 101; see also Morshed 2002). For the planner, the view revealed both the problem of and the solution to spatial and social disorder (Matless 1998; Mort 2004). Stepping stones to tie territories together was one thing, but aerial vision permitted a rather different kind of gaze; this gaze was abstract and it was targeted.

Aerial surveying and photography were enrolled as key 'weapons' in the colonial authority's ability to administer and manage their territory's inhabitants and resources. Along with the Ordnance Survey (Collier and Inkpen 2003), the Air Survey Committee, a sub-committee of the Colonial Survey Committee which acted as the right hand of British colonial mapping, was to be instrumental in this role and the development of techniques utilized in the Second World War (Collier 2006). The Colonial Survey Conference of 1927 and its subsequent discussion saw various military personnel as well as aviation experts come together to discuss the utility and potential of air survey in the empire. Aerial survey marked the application of professional knowledge and scientific expertise to warfare and colonial administration.[4]

At the conference, Sir Herbert Stanley articulated the vital role aerial survey would play in Northern Rhodesia. '[A] map is absolutely essential', he said, 'if we want to make roads or open up the country at all'.[5] In South America, opening up the country to the gaze and the reach of the colonial authorities required the initial penetration of dense forest paths that were invisible from below: '[O]ne can almost say with assurance that the almost impenetrable yet very valuable rubber forests of the Upper Amazon, to take one example, will never be successfully mapped until the services of the aeroplane are requisitioned.'[6] The aeroplane could be used, Stanley argued, to trace the river courses – their navigability, nature and direction – so enabling their subsequent penetration by boat.[7]

There is, however, another vantage point from which the aerial gaze has looked. Others present a far more complicated and diffuse picture than aerial survey. In fact, on the electronic battlefield and in the airport, 'the lived', which de Certeau saw was so alien to the aerial view, draws the aerial gaze down to more *grounded* bodily performances of mobility. The security of these pathways is dependent less and less on overarching visions – say from CCTV that monitors the terminal space – and relies on a more honed focus, trained on the bodily signatures of the population (Klauser 2009). The gaze with which we are familiar is a gaze of scrutiny.

Akin to the aerial photo, the airport gaze is a strategic one. The emergence of biometrics in contemporary airports, border zones, security spaces and everyday life sees the systematic use of biological and bodily data as a means to identify and manage risky mobile populations, focusing upon not territory but vectors (see previous chapter). The origins of biometrics are actually quite old. Even earlier than aerial survey, in the colonial context of the British Raj, which was engaged in all manners of surveying the land and its population, surfaced the pioneering technique of fingerprinting. Mobile and nomadic tribes consisting of habitual and deviant criminals made portable by the railway needed to be managed as a graspable population by methods that could differentiate one person from the next (Sengoopta 2003). The research of Francis Galton, Alphonse Bertillion and Henry Mead saw anthropometric techniques of identification give way to fingerprints in the wake of the Criminal Tribes Act of 1871. Previously indistinguishable subjects, now distinguished by the relative static pattern of their fingertips, could be enrolled by the new technology of fingerprinting into a searchable *population* of records against which traces could be measured, judged and verified. As with colonial survey, fingerprints could help solidify possession: through identification one could ascertain the identification of someone purchasing or selling land, drawing a pension or signing a contract.

One of the promises of the first biometric schemes and aerial photographic survey was therefore their capacity to appropriate new objects for an administrative gaze. The following section explores this as a process of *revealing*.

Revealing

Topographical mapping is a prime example of processes conducted by British intelligence operations in Palestine and Egypt during the First World War (Dowson 1921). For military reconnaissance, aerial photography had two main purposes, which we can interpret as a form of targeting (Finnegan 2006). The function of the air photograph was contested. The purpose of the photograph was not necessarily to 'show primarily *what* an object is, but to show *where* it is, and enable us to place it in its correct position on the map'. On the other hand, from a 'strictly intelligence point of view', it was suggested that the primary (Macleod 1919) function of a photo from the air was to record transition: 'what is on the ground at any moment'. Through successive photography the imagery could allow the monitoring of the enemy's organization, 'their natural change and development', and so 'detect and forecast his military plan' (Macleod 1919: 386). Aerial photography was primarily a technology of location (Kaplan 2006b).

Affording a perspective far different to that from the ground, the possibilities of the emerging field of aerial archaeology formed a new disciplinary science as pilots discussed their findings in places such as the Royal Geographical Society's journals. Obscured to the body on the ground, Colonel Beazeley pioneered archaeological interpretation during his flights over Palestine. He found that when riding an aeroplane *en route* over enemy territory, on the desert below him a set of outlines of a series of detached forts were revealed, 'whereas when walking over them on the ground no trace was visible'. To the terrestrial observer, these forms were invisible, merely evident as 'meaningless low mounds scattered here and there' (Beazeley 1919: 330).

In the burgeoning science of aerial archaeology, development went hand in hand with the recovery of the past. British officer Beazeley's surveys of Eski-Baghdad 'revealed the true character of what otherwise would have appeared as mere meaningless low mounds at that great scrub-covered ancient city site' (1919: 330). There was much to be gained from aerial archaeology as it gestured towards quite different land forms of far more strategic origin (Stein 1919). The interpreter's analysis of hidden enemy positions faced almost the same difficulty of identifying objects which had been concealed (Curwen 1929); for both, 'great care is required in the recognition of features' and both could be 'verified constantly by the actual examination of captured trenches' (Reeves 1936: 104).

The aerial archaeologist penetrated down into the layers of history by taking in a more complete picture of a whole. This was an apt metaphor for the military intelligence officer's interpretation and *uncovering* of hidden traces of evidence, or the colonial administrator's search for natural resources. Partially destroyed vegetation and parched ground was an obvious sign of oil ripe for development. The camera would 'reveal the areas which have this distinctive feature, the tributaries which run into the parent spring, and the forest roads and approaches which will be of use to us', all of which could be missed, as explorers on foot 'might pass near to an oil spring without seeing it in the dense jungle'.[8] In Palestine, aerial reconnaissance would provide useful and detailed maps of Gaza in a 1916 survey and the photographic reconnaissance of the Turkish system of trenches which was detailed by Captain Thomas in the *Geographical Journal* (Thomas 1920). Observers' values and judgements of what counted, looked interesting and matched their experience of shapes and patterns turned their vision of the ground into a chequered patina of past forms and processes (Brophy and Crowley 2005).

By 1929 the aeroplane was involved in considerable explorative projects. In an expedition to map the Antarctic and develop British territorial claims, Sir Hubert Wilkins described how his aerial survey pulled back the veil cast over the mysterious terrain of the frozen land-mass, enveloped in cloud and mist. Aerial survey allowed one landform to be seen in 'its *true* relation to another situated not far from it' (Wilkins 1929 cited in Mill 1929: 377, emphasis added). In both survey and exploration from the air, things could be seen '*in their structure*' (Roland Barthes in Deriu 2006: 206). The survey revealed truths, it exacted a struggle 'to decipher a multiplicity of fragmented signs and reconstitute them into a signifying whole' (Deriu 2006: 206).

In other contexts surveys enabled administrative authorities to fix down populations and resources to particular locations where they could be managed. It was to place the population within a wider colonial terrain of the logistics of government. T. E. Lawrence volubly pointed out the importance of aerial survey as an administrative and tax-gathering technique (see also Dowson 1921):

[T]he tax gatherer would record his crop measurements – a most important item – and various other departments would likewise gain. [...] Palestine requires land registration speedily and cheaply. Even now the quickest and cheapest method is by air photo survey combined with the necessary triangulation. [...] As for surveys in Australia, Canada, India, the air photographer and surveyor between them have a huge field in the Empire, and any country, such as China or Argentina, requiring land registration for taxation purposes or other mapping would save time and money by adopting the method. The present need is for further research, a small outlay, and co-ordination between surveyors and the Air Force.[9]

Considering the aerial surveying of the Niger river region, one member's submission to the 1927 conference encapsulates the modalities of governance afforded by the lens of aerial survey: 'We are beginning to tax the region', and yet, 'at the moment we do not know where the people are so that we may tax them'.[10]

The aerial survey devised order from chaos; survey divined truth from the lie of the urban malaise. The Eastern city was an especially problematic object for imperial ordering, being unplanned and as impenetrable as the jungle. The streets were 'narrow and tortuous', presenting the 'greatest difficulty to the ground surveyor', as witnessed in contemporary urban counter-insurgent operations. Making sense of the city through an Orientalist optic of what Mike Davis (2009) and Stephen Graham (2004) term as 'demonic urban places' versus the Western 'securitized' city, today's airpower theorists are reluctant without blueprints or intelligence to tackle an urban locale that is 'decreasingly unknowable since it is increasingly unplanned' (Davis 2009: 204). Almost a century earlier, the survey of Peshawar was proving 'already of great use in the successful rounding up of disaffected frontiersmen and Afghans'.[11] Aerial photography could enable the colonies to locate their populations, manage rights over the land and administer government. The aerial survey *showed* the houses, the streets and winding alleyways so that they could be managed and policed, just as in today's context of counter-insurgency.

Debates over the scientific management of forestry by aerial photography served as a useful metaphor for the purposes with which aerial surveying had begun to be understood. Resource and population were brought into equivalence (Biggs 2006; Cronin 2007; Hoag and Ohman 2008). Survey by air could turn the unruly mass of species on the ground into a calculable and therefore governable object (see Demeritt 2001; Scott 1998). Both forest and population were problems to be dealt with together by the same means. Investigation into the utility of aerial surveying in North Borneo stood as a testing-ground for the future possibilities of air survey in forest management and cadastral mapping. Earlier reports in the region had previously accorded forest management with the powers to cultivate civilization as a form of 'empire forestry' (Barton 2001). The process of industrial and scientific forest management would develop nature into something productive and profitable while meeting the needs and propensities of the local populations. On the principle of technological rationalization, species of nature and man could be formally and legally dominated (Bryant 1997).

The solution to the realities unveiled by the aerial position was the interplay of survey and resource use as a form of social improvement. One report on forest management stated how the productivist management of industrial clearance could allow the aboriginal Indians' normal mode of life and activities to occur between and within the patternings of the fellings. Moreover, this would improve and develop their seemingly backward practices that exhausted the soil of its nutrients: '[T]he women will find the ground cleared for them, all the labour of cutting down the forest is avoided. They can grow their crops and in return for finding the land ready cleared they can put in the new trees and weed them for a year.'[12] Only under 'proper control' would the forests be 'readily and cheaply restored'. Reports suggested that the successive dynamics of the fellings would even meet the desired tendencies of the local people: 'As the fellings move on, the Indian fields will follow them up, and the nomadic tendencies of the Aboriginal will be pandered to.' Through this process both aboriginal and forest would be *developed* together. The aboriginal Indians would 'gradually learn more settled habits', while the forestry methods 'would go far to leaving the land in the most productive condition to which it can be brought, after the devastation wrought by the axe has passed by'.[13]

In this brand of managed coexistence or 'welfare colonialism' (Scott 1998), both colonial authority and aboriginal populace could accommodate one another. They would overcome the natural and 'intense struggle for existence between different species' of tree that could result in the 'disappearance of that species in favour of a rival species', or, in their lack of competition, homogeneous low-quality trees and potentially low-quality populations. Colonist and colonial subject could live together. The airborne camera overcame 'one of the most serious obstacles which had hitherto delayed the development of many of the natural sources of the Dominions and Colonies'.[14] It began to be seen as the first step to progress in the colonies and the initial task for any development activity. As Colonel H. L. Crosthwaite declared to the Royal Society of Arts, 'Surveys and maps should precede development and not follow it, if it is to be efficiently and successfully

carried out.'[15] Aerial photography 'can assist towards the scientific development of the Empire', he concluded.

Contemporary biometrics at the airport border enact a closer gaze of the
body as a territory to be captured and managed. Unlike colonial administrative power, which the aeroplane anchored through the identification of
land and population, 'biometrics derives from the human body being seen
as an indisputable anchor to which data can be safely secured [...] the body
as a source of absolute identification' (Amoore 2006: 342). Biometrics
reveal information about passengers that allows them to be identified.

Exit–Entry systems such as the United States Visitor and Immigrant
Status Indicator Technology (US-VISIT) programme inaugurated in 2004
enact a totalizing vision, much like the aerial photographic survey. They
capture parts of passenger bodies, and, like the tessellation of the aerial
photo, add up to a bigger image of a population. Making efforts not to discriminate between passengers, the programme fingerprints and photographs passengers 'indiscriminately'. These include visitors from both
non-visa and visa-waiver programmes who intend to stay under and over
three months. As Barbara Epstein has it, 'discrimination' has been replaced
with "enrollment" (2007: 159). In other words, biometrics offers a familiar
stepping back or process of removal performed by the aerial photo. It is to
understand the fingerprint imaged, or iris presented, in the wider context of
a population of records.

Like the aerial photo, the biometric airport/border is able to 'foresee' in
order to pre-empt or prevent certain futures: these might include the risk of
terrorism, unwelcome migrants seeking residence, or drug smugglers. The
biometric border does this by using imagery of the body in order to locate
who the passenger is. And as with the aerial camera, the technological methods of completing this task are many. In Canada, Colleen Bell notes how
vision detection strategies are employed through a range of 'border security
enhancements' such as 'LiveScan digital fingerprinting [...], an RCMP Real
Time Identification project that enables the electronic recording of fingerprints for instant verification, and biometrically enabled smart chips that
use facial recognition technologies "to interrupt the flow of high-risk travellers"' (cited in Bell 2006: 159).

As a magic bullet, biometrics is the technological foundation of progress in
airport security and the management and securitization of mobilities across
borders (Lyon 2003). Airport biometrics, like the aerial survey and photography, constructs strategized spaces. These spaces do not hold or capture
populations within them but add up to pictures of individuals who make up
larger populations of mobile people. The biometric airport terminal conjoins
both body and image. The passenger-subject is recognized as a sign – a
number facilitating the airport's function as a 'machine of representation that
produces and reproduces the subject as a representation, as a "word without
body"' (Diken and Laustsen 2006: 450). For Bulent Diken and Carsten

Laustsen, the airport resembles a 'zoopolis', where 'citizens' are 'reduced to naked bodies because the biometric technologies of surveillance can only "scan" and recognize the subject as a body or body parts' (2006: 450).

Systems such as US-VISIT perform as tools which manage populations through verification and identification. With this information, 'decision makers' are able to judge whether a passenger should be allowed through or halted and returned. Before entry and in the event of it, biometric programmes such as US-VISIT enable Consular Officers to collect biometric along with biographic data, whilst passengers under visa-waiver programmes submit their details for the first time. These are both a means to establishing identity. Immigrant biometric schemes have been used such as the Eurodac, which stores the fingerprints of all asylum seekers across the European Union in order for states to identify a passenger's identity (Van der Ploeg 1999, 2003).

On the other hand, returning visa-waiver passengers and passengers who have gained visas have their identities *verified*. Palm recognition and now iris recognition technologies are deployed in many instances of facilitating airport priority passengers and frequent flyers who are pre-enrolled. In airports the Privium frequent flyer programme at Schiphol in the Netherlands as well as Canada's CANSPASS (Muller 2004) and the US 'Trusted Traveller' are cases in point of the elite use of systems where trusted and frequent flyers have opted in (Adey 2004a; Cresswell 2006a; Salter 2004).

In other words to *know*, and *know* accurately – knowing implying a degree of confidence over the identity that has been captured in order to be later verified – requires the ability to see or capture those biodata accurately. Airport biometric technology has continuously undergone questioning for its ability to determine one identity from the next. In much the same way, debates surrounding aerial survey have always orbited around the issue of discernibility. Could an aerial photo be used to differentiate between targets, trees, rivers, streets and populations? Colonel Harold Winterbotham, in charge of the Ordnance Survey, stated the problem incredibly simply: '[Y]ou cannot survey unless you can see.'[16] By the late 1920s, discussion occurred over just this question: was aerial survey good enough to *see* the wood for the trees? It was a question of resolution and the ability to differentiate from that resolution.

Resolution

In a 1927 report investigating the potential for the stock-mapping of the forests of British Guiana, the Assistant Conservator of Forests arrived at the conclusion that '[a]erial stock-mapping [...] is impossible', primarily because 'little or no differentiation between trees growth was to be seen'. Even creeks were entirely invisible.[17] At a height of around 7,400 feet with excellent visibility, the Borneo conservators conducted experimental flights to ascertain the viability of aerial survey in forestry valuation. Both assistant and conservator made reports in order to submit a fair recount of their experience from each side of the aircraft. Unfortunately, from either side of the plane the results were dismal. As their report stated: '[I]t was found impossible to distinguish between different species except in the case of one tree, so scattered as to be rare, and which appeared to be in flower, each individual tree showing up very clearly as a yellow spot in the surrounding mass of green.'[18] Even with the support of a stock map formulated from already surveyed forest, attempts to read the timber supplies failed. It 'proved quite impossible to distinguish what the prevailing type of forest was, and it is clear that no stock-mapping from the air will ever be possible'. Mainly because the varieties of species that composed the forest mass were of relatively similar crowns and foliage, the report re-emphasized its conclusions: 'It will never be possible to stock-map the evergreen mixed forest of the Colony from the air.'[19]

The poor results of the experiments drew discussion at the 1927 survey conference. Although the Borneo example as well as efforts in Johore, Malaysia, had been 'extreme' cases of mixed forest species, they questioned the future utility of aerial photography as a surveying tool. For Colonel Winterbotham of the Ordnance Survey, the findings in Johore had found vegetation 'so dense that you cannot see anything through it'. Trees and foliage made judgements of relative heights for contouring virtually impossible given the error margin possible of around 100 feet. The practicality of aerial photography as a useful technique for surveying was even cast into doubt given that its use might only lie in identifying suitable areas for ground survey, 'a question' which for Winterbotham 'does not quite fall within the scope of the term "survey"'.[20]

For those more in favour of aerial survey, differentiation between species was not the main issue, indeed E. M. Dowson (1921), Director of the Survey of Egypt, had earlier reported the difficulties of crop identification in Egypt. Aerial survey could give a more general idea of presence, location and practical extraction. With stereoscopic advances, 'while you may not be able to identify individual species in a forest, you can spot gregarious species and you can probably deduce the presence of a species'. For example, taking the case of mahogany, one proponent argued, 'you could probably recognise from the air the type of forest in which you are likely to find mahogany. You can also find out, and this is our great difficulty, how to get there and how to get the mahogany out when you have found it.'[21] In this sense, aerial photography presented not the resolution for individual speciation, but the identification of areas of forest and non-forest, as well as the lie of the land and its propensity to support ground survey activities and logistics (Figure 4.1).

By August, another valuation survey was underway in Borneo which was intended to identify where particular species might lay with the aim of

Figure 4.1   The aerial grid. (Reproduced with permission of the RSA.)

sending ground-based valuation surveys to those areas.[22] Finding little more than swamp, smaller patches of isolated forest and a further 3,300 square miles of low-value timber, the report was considered a 'great success'. Aerial reconnaissance, the report argued, 'in this instance has undoubtedly saved expense in Forest-Valuation by showing where *not* to begin surveys'.[23] It was this view which began to win out. Aeroplane photographic companies went to considerable expense to prove and lobby for the application of air surveys, especially in the forestry context. By 1928, more aerial surveys were in the progress of development with several applications being made to the Colonial Office.

Following these reports up, Mr R. Bourne, a scientist active in the Canadian Forestry Service, made the strongest case for aerial survey. For Bourne, although an 'air photograph would not penetrate' the likes of dense forest undergrowth, it 'indicated very clearly the distribution of the different vegetation types' (1928: 19). The photographs gave the observer a very good idea of the distribution of the vegetation and an indication of 'how the forest could be developed' (1928: 19). But more than that, the aerial view enabled one to see trees and timber in relation to other things, in a way, Bourne argued, that came to stand for the cooperation and 'future prosperity of the British Empire', which depended on its exploration of the forest, agricultural and mineral resources of the Colonies, India and the Dominions (1928: 6). If these natural resources could be seen in relation to one another,[24] from 'exploration, general reconnaissance and preliminary survey; forest, and agricultural survey and land registration; hydrographic and irrigation survey; geological, mountain and railway surveys; archaeological survey' and map-making expeditions, their future effectiveness would be secured.

Invoking the metaphor once more, Bourne argued, 'Literally the forest officer often cannot see the wood for the trees', the difficulty lying in the 'the survey and mapping of areas intervening between the lines traversed' (1928: 5). And yet, air survey enabled the visioning of trees in relation to their soils, to the geological features of the region. Encouraging the cooperation of ecologists, soil scientists, foresters and airmen, aerial survey could correlate the patternings and dynamics of the relation between these different forms of disciplinary expertise and the material phenomena they wanted to understand. Once accomplished, zones of favourable soil would be understood next to zones of forest to be reserved for commercial exploitation or protection. A correlation between 'fly', 'climate', 'geology', 'soil', 'vegetation', 'fire', 'game' and 'man' would be brought into sympathy and alignment, facilitating the further development and the tapping of productive areas. The 'local government', Bourne wrote, should be 'in a position to view in real perspective their present and future problems' (1928: 16). The aerial view shook things up. Like a sieve, its perspective filtered out the noise and colour to reveal a more meaningful image that could be harnessed. The distant view of the camera in the air, summarized the Royal Society's chair as they debated the subject, 'was necessary to convert chaos into order'.[25]

As with aerial survey, the biggest controversy to face airport biometrics has been its efficacy. Usable biometrics at the airport terminal depend upon the capacity of that system to be able accurately to identify or verify across an enormous population of travellers, a seemingly chaotic multiplicity (see Agre on these issues, 1994, 2001). In order to do that, enough information about an identity must be capturable or recordable. This must occur in order for that system to be sure one person is not confused with another. A biometrics system without the necessary resolution increases the propensity for so-called 'false positives' – when someone is incorrectly identified as someone else – and 'false negatives' – when an apparent threat is not identified and slips through the security net. Given the enormous numbers of passengers, just a small percentage of false positives could lead to thousands of passengers incorrectly identified. As Lucas Introna and David Wood note, during the first few years of a test at Palm Beach Airport, Florida, 'the system achieved a mere 47% correct identifications of a group of 15 volunteers using a database of 250 images' (2005: 82). These false positives are now a much lower percentage. However, given the millions of passengers enrolled within such systems, the biggest issue facing biometrics is how to ensure resolution of detail is sufficient in order that verification can be completed with confidence.

Airport border biometric programmes have required a number of different bodily indicators from which to capture reliable and accurate bodily information. The US-VISIT programme initially ordered passengers to submit the fingerprints of their two index fingers and an image of their face; by 2009 the number had increased to all 10 fingers. Others rely upon iris recognition, and some use the prints of the hand's palm. The object is again not necessarily about ascertaining *what* an individual is, but rather it is about locating their details within a schema of risk and trust.

In the same manner as the earliest forms of aerial forestry survey, certainty over what a member of a population is *not* has become a logic employed to filter out intended targets and the allocation of resources. Pre-enrolment of passengers within biometric frequent flyer programmes, of which there are many, works by building up enough information about a passenger in order to suspect them of *not* being a terrorist – or, to be more precise, to allocate them a lower category of risk. By this technique, more intense forms of scrutiny and interrogation can be targeted towards the locations of the population more likely to be carrying a bomb, a biological weapon or drugs. The resolution of the survey and biometric security is necessary, and the fidelity to their objects is almost as important.[26] And yet, in order for that information to be useful, the aerial gaze has needed to produce knowledge which is as mobile as the objects of its vision.

Circulation

A simple image such as an aerial photo or a biometric print is not enough in order to manage susceptible, sedentary or mobile populations. Rather, these images need to be made commensurable in order that they can be enrolled into wider systems of equivalent knowledge where they may be judged, valued and circulated.

Making maps from aerial survey photography was a complicated process that required the labour of several military officers, interpreters and cartographers. For the first RAF reconnaissance flights in Palestine, the process would usually involve two copies of every photo taken. One contact print would come direct from the RAF, which was received clean, or before it was placed or keyed up by a photographic officer, who would have a team of around four men underneath him. This would normally be provided some time after the flight (around two to four hours). The second copy would have been provided by the photo officer, who marked the enlargement to show all things of military importance and would be normally taken a day afterwards.[27] During the Second World War, this process became increasingly

Figure 4.2   The photomosaics are combined and turned into a map. (Reproduced with permission of the RSA.)

swift for air-photo reconnaissance and interpretation – a process which emerged into the sequences of the 'kill-chain', the progression from targeting to destruction which is today being increasingly compressed (Graham 2006; Gregory 2009).

Contacts would be examined and divided into groups for sticking down. These would normally consist of 10 to 20 photos of the same scale and size, with identifiable points at each end for scaling and overlapping. Where an overlap was difficult between photos, it would be marked with ink. The process created a continuous stream of images. From stereoscopic examination, lines could be used to mark in the form of relief of the country captured. Colours were used for marking up groups and objects of particular interest. Things of military importance were marked in red, form lines and houses in brown, with trees in green. Pantographing and compiling would then see tracings which followed the outline and detail of the photo (Figure 4.2). The result of all of this: to break up a near featureless landscape populated by indistinct landforms into the rational space of the grid. Vegetation, landforms and populations are set against what Blomley describes as 'an inert structure. Space is marked and divided into places where people are put' (2003: 128).

Rather than dealing with the corporeal, unruly flesh-and-bones body, biometrics breaks the body down into data in order to give the body its equivalent properties with others. This is a means to tether that body to representations of the subject's identity and an archive of their mobility. The movement of the body's information within this architecture has required the reminiscent dissection and tessellation of the body's image into workable data that are interoperable and therefore mobile.

Tying together an assemblage of security systems (Haggerty and Ericson 2000), the United States Enhanced Border Security Act of 2002 lay down the foundations for the integration of some 20 systems and databases upon which US-VISIT functions. These include systems such as the Arrival Departure Information System (ADIS); the Advance Passenger Information System (APIS); the Computer Linked Application Information Management System 3 (CLAIMS 3); and the Interagency Border Inspection System (IBIS). IBIS is further connected to the Interpol and National Crime Information Center (NCIC) databases, while the Automated Biometric Identification System (IDENT) stores biometric data of foreign visitors, whereas the Student Exchange Visitor Information System (SEVIS) contains information on foreign students in the United States. Biometric information is transmitted across and between various databases of other law enforcement agencies, immigration and justice departments.

In aerial survey these terms were understood as another manner of *cooperation*. The break up of the land into the logic of the grid effectively liquefied the immutable properties of space into a region of transaction and exchange. In Palestine, photography was mulled over as an effective means to conduct cadastral surveying or at least prepare the way for it. The surveys were contemplated as a means to remove the 'uncertainty of land tithes', impose taxation and deliver the 'possibility of transferring waste and State land to Zionists' (Gavish 2005: 41–2). Colonist logics of grid imposition fixed the identity of land and its ownership not so its provenance could remain static, but in order for its identity and properties to be circulated and exploited (Blomley 2003).

By 1929 more photographic commissions were underway, emphasized in their ability to undertake survey rapidly and cheaply, with the result of re-inscribing space in a manner that it could be reorganized and leased. Whilst immediate revenue from the surveys was unlikely, the advantages were seen to lie in further settlement of the area by Europeans, and in the developmental aims of improving 'lines of communication for natives seeking work in the industrial and European farming areas and in the opening up of additional markets for native-grown food'.[28]

Both contemporary aerial surveillance and airport biometrics are far less labour-hungry or time-consuming. A signifier of character in Chapter Two, the face is an obvious marker of identity and necessary for mobility across borders (Salter 2003; Torpey 2000). The face and other parts of the body function as one's passport (Van der Ploeg 1999) as the body must then show up, be present, ready, corporeal, to be read as various thresholds are surpassed (Agamben 1998). From fingerprint and iris recognition technologies to the simple photograph imprinted on one's passport, the unity of the body is undone by focusing in on pieces of it. These pieces stand for the whole. Pattern recognition filters take faces or fingerprints as their object of capture. Like the aerial photo, exactly how the face or fingerprints are stored, transported and scrutinized sees their features transformed into a territory or resource, requiring similar tolerances of capture under the remit of *interoperability*.

The International Civil Aviation Organization's (ICAO) efforts to standardize travel documents have developed the hallmark of facial recognition technologies regulating how passengers' faces are captured in order that the information can be transported and compared across geographical distance and between states. Like the rules of the aerial photo, the ICAO's Globally Interoperable Biometric Standards ensure that the face must be captured from the correct distance, cut in the right proportions and stored in the correct manner so that the image is recognizable by other border agencies encountering the passenger. Numerous drafting groups have legislated for particular tolerances in head roll and tilt to plus or minus 8 degrees, and other definitions of head size and position. These included defining the parameters of the horizontal and vertical positions of face, the width and length of the head, and another set of perameters for children under the age of 12.[29]

For example, ICAO photographic working group guidelines state that a 'portrait must be a centered, frontal view of the full face', they specify the focus of the image 'from crown to chin, with both eyes open', whereas the 'head height must comprise 70 to 80 percent of the portrait height'.[30] Careful lighting is necessary to expose the face in the right proportions: '[A]ppropriate illumination techniques shall be employed and illumination used to achieve natural skin tones and a high level of detail, and minimize shadows, hot spots, red eye, and reflections, such as sometimes caused by spectacles).' Furthermore, facial comportment and expression are increasingly important in facial recognition. Pose should be 'neutral and non-smiling, with both eyes open and mouth shut' (Figure 4.3). As we will see with the necessity of steady flying in relation to the ground which was conducive to successful aerial surveying, 'the rotation of the subject's head must be less than five degrees in any direction (roll, pitch or yaw)'.[31] Although 'roll (or tilt)' can be corrected later using image processing software, 'similar post-acquisition correction of pitch and yaw are error prone and could introduce substantial distortion to the image'.[32] Bodily disposition, in other words, is vitally important in the construction of the aerial view.

Exactly how much of the face is displayed on passports and in the electronic record is just as vital. So that they may be turned into what Gates calls 'stable, mobile and combinable information' (2006: 430), while saving on storage, it was contemplated that images should show just the eye/nose/mouth features. However, the working group decided that 'the ability for a human to easily verify that image is of the same person in front of them, or

Figure 4.3   Representations of the face in ICAO guidelines. (Annex A, Photographic Guidelines, http://www.icao.int/. Reproduced with permission of the ICAO.)

appearing as the photograph in the data page of the passport, is diminished significantly' when these features are not pictured in relation to the rest of the head.[33] Again, being able to distinguish the characteristics of subjects or tree species necessitates their visioning in relation to a whole. Even the exact size of image, in terms of the bytes of data the photograph takes up, is standardized by the ICAO in order to ensure that the quality of the image is good enough to recognise, whilst it is small enough to optimise data-processing times.[34]

The biometric and the aerial photograph are problematic in the sense that because of their fidelity to the object, body or landscape, they capture space as it was at a particular moment in time. The securitization of such biometric 'bodies-information' ineluctably becomes a process that fails to account for the transformation of those codes through disease, genetic alteration, and body-subjects' *in-formation* (Dillon 2007b: 25). On the other hand, Epstein explains how biometrics, in its focus upon the 'measurement of life', 'rings of biopower'. For her, 'it is functionally concerned with the life of these bodies in that it needs to ensure these are "alive" for the system to be optimized, as one biometric manual emphasizes' (Epstein 2007: 152). A body part such as an iris is therefore considered the most reliable of biodata because it is the hardest to overcome through fakery: 'Iris recognition technology requires that an iris is alive in order to work' (Bolle et al., cited in Epstein 2007: 152).

In aerial photography, because of changes inherent to the life of the environment they attempted to capture – the regular sequences of seasons and their alterations on the landscape, the daily changes of the sun's position upon 'the orientation of shadows', the transportation and farming routines, the arrival or non-arrival of a train – 'a picture, pure and simple' was deemed unsuitable. 'Convention must be utilised' in order to portray becoming.[35] Conventions of capture and representation meant the knowledge portrayed within a photo or an aerial map could be understood across borders, countries, and across expertise. 'Conventional signs' might include the use of shadows orientated from a common direction, the sun pictured from the south being 'perhaps the most suitable', assisting the orientation of the map. Shadows were important because they would enable the map's reader to differentiate between objects pictured, giving an idea of upstanding or sunken features on the map.[36]

Through these processes the aerial photo creates something that can be got at. Of course they are made precisely so one can know, touch and organize their assets (territory, subjects), as well as others. However, in subjecting populations to administration, warfare or biometric management, new subjects capable of being administered, seeing and being seen in a particular way, are created. I want to suggest that attempting to subject populations to the aerial view has required the simultaneous subjectification and disciplination of populations and bodies to those processes.

Techniques of the Observer/Observed

First summarized in the British Air Survey Committee's report of 1923, the issue of how pilots should fly in order to develop the best and most accurate aerial photographs for survey drew upon the experience and expertise gained in Palestine and Egypt. The particular terrain of the Palestinian region, a 'steep and rugged type of country' which, while mountainous, was free from the friction of anti-aircraft and hostile machines, provided an ideal testing-ground. The airmen could 'take photographs level and free from the distortions due to tilt and change of height' (Tymms 1927: 6); in other theatres of war 'the flying of the plane was never steady enough to guarantee anything but distorted photographs'.[37] The desert terrain didn't necessarily soften the pilots and observers to the task, but rather permitted the time to practise and hone their skills. As a result the airmen were 'more expert in taking photographs for mapping purposes than were those on the Western Front, whilst Survey Units had more experience in eliminating the effects of tilt on the Western Front than was the case in Palestine'.[38] The particular terrain made accurate surveying and eventually map-making extremely difficult.[39]

Without these kinds of conditions, the pilot had to work hard to avoid 'bad flying' and the subsequent errors and distortions their flying would imprint onto the photographic plates. The grid of photographs and maps (as shown earlier) was made possible by steady and controlled flying.[40] Figure 4.4, reproduced in the Air Survey Committee report of 1923, illustrates the difficulty facing the pilot in producing accurate and usable photographs; the pilot was forced to conform to the grid.

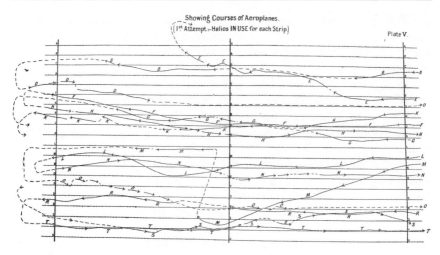

Figure 4.4   Gridding the aeroplane's path. (Crown Copyright.)

To capture the view from the air therefore required very particular kinds of subjects, trained and able to move their aircraft in the correct way, to coordinate in sympathy with the terrain of the ground below, the observers they worked in tandem with, and with interpreters able to think and *see* in the desired fashion (Tymms 1927).[41] The grid imposed on the ground below was, thus, imposed onto the pilot themselves, the lines of the grid becoming the walls to the pilot's trajectories.

Double exposures, camera jams, underexposures, deviation from the courses followed, gaps in the photos and the pilot's inability to locate the aircraft's position – all these variables might mean photo distortion, and require the course to be redone, a sequences of photos re-taken and considerable work by the interpreter to adjust or 'fix' the imagery.

Accurate photography also necessitated the pilot's partnership in sympathy with his observing passenger, for which '[t]horough cooperation and a complete understanding are necessary' (Tymms 1927: 6). Without hostile aircraft to disturb the observer or the pilot, the working of the camera would be carried out by the observer, with the pilot focusing all his attention on the desired course and on keeping the aircraft as level as was feasibly possible. In harmony with the pilot, the observer would work to make subtle alterations to the levelling of the camera, making the exposures at the correct intervals and changing the boxes of plates.

Automatic cameras would remove some of the 'constant strain' (Tymms 1927: 6) placed on the observer continuously to make exposures, which was regarded as a generally tedious task, and was liable to result in breaks in the photos. Other technologies would facilitate the pilot's attunement to the level of the aircraft. Like the photographic instructions the ICAO imposes upon the generic photography of passengers' faces – including the yaw and tilt of the face, and other tolerances of distance and passenger position – the pilot and observer were disciplined by similar margins of error (Crary 1999). A simple spirit level was installed which the pilot would use to guide him. A more involved system similar to those used later in aerial bombing was to draw a table calculating the time intervals, with different ground speeds for exposures. Bombsights were suitable to perform this task, as time-distance and speed could target the correct time and place for an exposure to be taken. Either way, the pilot would have to make 'every effort to maintain a constant level at whatever height is finally determined' (Tymms 1927: 10).

Tilt finders were developed which made the distorting effects of tilt on the image easier to calculate and therefore rectify. The angle at which the photograph was taken could be known by the photo interpreter. Automatic cameras which recorded the height of the aeroplane, date and time, serial number as well as angle and tilt and direction of the optical axis all improved the possibilities for the interpreter to figure out the aircraft's position in space. Whilst different methods of computation or trial-and-error adjustment of the camera did not appear 'suited to our national habit' (Air Survey Committee 1923: 17), aerial photography and surveying involved the subject, pilot, observer or interpreter in a variety of technologies of discipline, orientation and correction.

The air-photo interpreters, clearly at the margin of a complex interplay between their own practices and the performance and style of the observer and the pilot, were no less subject to the same procedural rigour and training. In fact their role was even more markedly fixed and disciplined by the use of equipment which required a set of appropriate sequences of actions to use them correctly. So as to discover the 'elements' of an image and not to 'waste time or effort', absolute procedural rules were suggested for the progressive adjustments and alignments of hands, eyes, photo and the various moving parts of a machine such as the photogoniometer, the Fourcade-designed system that would structure the observer's sight in order to turn two two-dimensional images into a three-dimensional or converged one.

One instruction manual advised detailed movements and instructions for the operator's body:

> 1) Two positives are placed in the holders of the photogoniometers and rotated until the axis of tilt in each is horizontal. Each holder is then inclined at the previously determined vertical angle and clamped. 2) Set the distance bar so that the reading of the slide attached to the right-hand arm and carrying the pencil shows the length of the stereoscopic base at the required scale. 3) Releasing the clamps of the azimuthal levers, rotate the two photo carriers about their vertical axes until a stereoscopic view is obtained and the pointer is laid both as to the line and distance on a trig. Or other fixed point. 4) Rotate

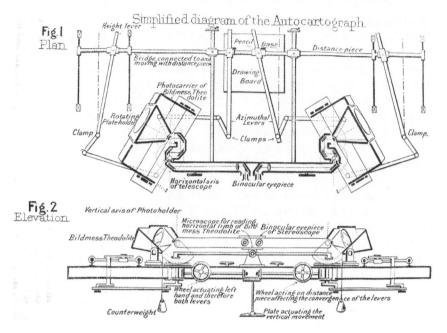

Figure 4.5   The autocartograph. (Crown Copyright.)

the horizontal scales of the photogoniometers until the required – previously calculated – readings or bearings are shown by them. These two scales are read together through an eye-piece placed below the binocular. They appear side by side and must be rotated until the base angle at one end of the base and that the other appear on the scale exactly opposite to each other. 5) Set the vertical arms or levers to read the correct height of the trig. Point on the bridge. 6) Tighten all clamps. (Hotine 1927: 58)

The photogoniometer's resemblance to the sorts of visioning devices Jonathan Crary (1990, 1999) has described convinces us of the subject's construction by the apparatus. Just like the airshow architectures, the photogoniometer forces the subject to it. It *must* be rotated in the right way. The arms and levers should be set, clamps will be tightened and a form of self-governance develops. Other visioning devices included the autocartograph (Figure 4.5).

The sorts of technological apparatus required by airport security screeners to conduct the biometric capture of passengers enact a similar relation between body and machine. More pertinently, however, is the way airport biometrics devolves its responsibility to the passenger themselves. Foucault's (1978) confessionary complex, which Mark Salter (2006) argues so typifies the border, is here enacted through a technological disciplination of the practices that disclose one's own body.

The increasingly automated character of such systems means that it is the passenger who operates as the 'user' or 'client' of airport biometric technologies. The regulation of mobilities across and over borders, therefore, occurs within the airport/border itself. For Louise Amoore, 'the trusted traveller' specifically is a prime object of self-regulation techniques:

> The US VISIT in-flight video has an animated Tom Ridge warning that the traveller has the responsibility to record their own electronic fingerprint at exit kiosks in the departure lounges. Rather as a credit rating is derived from past patterns of responsible financial borrowing, the trusted traveller is the individual who governs his own mobility and establishes a low risk mobility rating. (Amoore 2006: 343)

Nowhere is the responsibility more clear when US-VISIT is marketed by the Department of Homeland Security (DHS) as 'US-VISIT and You'. The TSA's 'checkpoint evolution' (discussed in Chapter Five) uses a similar strapline: 'Terrorists Evolve. Threats Evolve. Security Must Stay Ahead. You Play A Part'.[42] But the passenger's autonomy is shaped and curtailed. Airport biometric visioning technologies subject their users to certain routines or structures of action in order that the machines are used in the *right* way. US-VISIT first required that passengers 'checked-out' of the US by using a self-service kiosk system before they were to board their flight. Design of the kiosk software and material architecture ensured a natural and secure feeling of passage through each stage of authentication. By ergonomics, the passenger is encouraged to place their hand in the correct position; the system ergonomically 'forces' 'the hand to be in the best position to gather the highest quality of fingerprints presentable by the user'.[43] Indeed, the ICAO's tolerances of capture are familiar and habitualized by many passengers now used to displaying their face, or presenting their fingerprint.

Ergonomically designed hand readers that smooth the entrance of passengers' hands into a compartment adequate to the machine are therefore now up-scaled into the function of kiosks that seek to adjust the passenger's orientation towards their own scrutiny. The SmartGate system in Australia lets passengers with an ePassport do the checking in themselves through Passport Control. A kiosk, as Dean Wilson and Leanne Weber describe, 'takes an image of the traveller's face and matches that against the holder's passport image' (2008: 134). In many respects the border process is self-disciplining, reliant upon the semi-automatic character of the technology of biometric scanning and the bodily performances of the passengers using them. In other examples, kiosks or booths lock passengers in until they have completed their authentification processes. Physical barriers such as turnstiles and other sorts of controlled portals enable access once authentication is given.

This kind of lock-in is expressed in larger movements through security screening as the space of the security checkpoint is zoned and planned to compel correct and efficient movements. Simple signage and intuitive designs and processes are created to afford the right kind of screening flow and more compliant passengers. Pre-screening preparation and instruction zones designate passengers for their future instructions. Floor materials and colours mark areas intended for queuing whilst other colours and materials intuitively designate a barrier. What the TSA call other 'material, spatial, or lighting clues' are used to 'reinforce the appropriate paths' throughout the security areas. With the signage and the design, passengers or 'users' are directed and instructed, 'increasing the speed and perception of service'.[44] As Epstein has it, through these kinds of processes, '[p]ower is felt, quite literally, right at the fingertips; the experience is of an encounter with power, of a new kind: immediate and sensory, and yet harmless at least physically' (2007: 150).

Aerial survey could develop and train pilots and their crew (observer) in various aeronautical skills and qualities. This was particularly the case for character. In colonist discourse, survey appeared to develop the character and capacities of the subjects who learnt and practised its techniques.[45] Construed as a civilizing act, the ground geodetic and cadastral surveying of West Africa required the right kind of African native, who would be trained and civilized through their tuition. The setting up of a survey school in Nigeria found the 'African native' to be 'very quick and adaptable', but not highly educated. It was found that the 'chief difficulty' was 'inculcating some standards of honesty and firmness of character'.[46] The ground survey work of the bush was particularly suitable 'for the employment of natives', who, if supervised and trained by a European, could produce an 'astonishing output' from their character training. African or Arab natives would be given no such training or purpose in aerial surveying as it was the RAF officers who would benefit. T. E. Lawrence was vocal in advocating the aerial survey as a form of training. In an article published in the *Daily Telegraph* he argued that survey would give the personnel of the Air Force, 'who are young, keen, and only too anxious for some useful work, a practical and definite object in flying with certain conditions to fill'. Air survey, more so than 'depressing' practice flights, would have enormous practical value.[47]

Hamshaw Thomas advocated a programme of survey work. In order to train pilots and observers in reconnaissance and photography, where it was not possible to be employed on active duties within definite 'national, military, economic and Air Force points of view', an annual series of air surveys would be of 'great value for both retaining and for results'.[48] Pilots and observers, he suggested, would 'find themselves in new country' and they would learn and become familiar with that country in a short period of time. They would develop skills in picking out 'suitable landmarks, and would have to fly carefully and accurately in carrying out the work'.[49] What's more, echoing Lawrence's argument, they would feel that work of 'a tangible and permanently useful character was being accomplished', thereby increasing their motivation and keenness to obtain good results. It would, moreover, justify more than the utility of the Air Force not only in war-time but also in peace, securing a stronger position 'to appeal for an appropriation of public funds'.[50]

The character-building qualities of the technological citizenship afforded to the employees of the aerial survey is reinforced in the biometric enrolment of passengers not only into a system of border regimes as citizenship rights, but also into a performed system of confession and belonging (Salter 2006). To a degree, argues Mika Aaltola, 'a person realizes his or her own identity and respective place in the prevailing world order through the pace of the walk through an airport' (2005: 274). The bodily performances required to submit one's fingerprint, iris scan or facial photograph enact an 'imperial pedagogy' of learned submission to the airport's security regime: 'The "stops" signify the presence of a higher authority while the "goes" indicate a sense that one is authorized, deposited with authority' (2005: 274).

Again, US-VISIT's 'biometrics and you' places the onus and responsibility for security upon the consumer, just as their efforts to enhance or evolve the 'check-point' supposedly create an experience that 'you own'.[51] Your border – a border for you – the investment made is that by complying with US-VISIT or a biometric scan is to play your part in the wider security of the airport and the nation.

Three-Dimensional Vision

We know how the aerial gaze is many; it is multiplied and situated in different contexts. It is also a vision that is practised and touched. It is not simply ocular or visual, but an assembly of practices and materials – a combining or, more appropriately, *convergence* of information from two or more different perspectives. Louise Amoore and Alex Hall (2009) place their examination of bodily x-ray backscatter technologies that enable airport security officers to see through clothing in the context of much older genealogies of seeing and touch within medical discourse. In contrast, the painstaking dissections and resections to occur on the table in this context were not by the surgeon, but by the interpreter and the cartographer's hands. The aerial photo is reduced to parts and sections and resected to coordinates. The whole may be observed, whilst parts of that whole are dissected into parts of a grid.

Eyal Weizman's (2002) groundbreaking critique of Israel's politics of verticality has unpacked the contemporary uses of this kind of convergence in the process of what he has termed 'hologrammatization'. Through the same processes of stereoscopy we shall explore, this reveals 'higher and lower buildings […] hills, mountains and valleys'. The 'specialized scrutinizing gaze of the analyst' therefore enables the transforming of 'the two-dimensional prints into a three-dimensional simulation'. By these projections, the analyst can begin 'carefully to identify targets or precisely assess the impact of previous raids' (Weizman 2002). These strategies are targeted towards the increasing regularity of the urban in war (Coward 2006). This is an urban addressed through its 'sheer three-dimensional complexity and scale', 'three-dimensional entanglements' which undermine 'expensively assembled and hegemonic advantages in surveillance, targeting and killing through "precise" air- and space-based weapons systems' (Graham 2010 in press). Surgical cuts and strikes at counter-insurgents (Gregory 2008) can only be fought through other three-dimensional counter-strategies of visualization.

The ability to see stereoscopically has almost always lain at the heart of aerial photographic techniques for purposes of both survey and military reconnaissance. To be able to see stereoscopically required the use of two images taken from the aeroplane in order to vision not a flat image, but a three-dimensional gaze that enables its observer/constructor to perceive depth and scale. As before, the ability to generate an image of this kind required close accommodation of interpreter and technology; criticisms abounded that it even restricted flying (Hotine 1927). Certain equipment was developed not because stereoscopy was necessarily complicated or mechanical, but because it required a peculiar level of control over how one was able to focus upon the detail of the images. Assistance was provided from over-engineered technological equipment which addressed the human's systems, structures and anatomy of optical sensibility.

Stereoscopy essentially needed two images taken from slightly different positions of separation of the same object. This meant that an aircraft would need to take two relatively quickly taken images in a strip. If both photos were viewed at the same time and positioned at the right distances from each other from the eyes, they could be fused into an image with the characteristics of three-dimensional form. With this information, the interpreter could begin to pick out the peaks and troughs – the relief of the land – radically transforming what a photo could convey. Similarly to the photogoniometer, the cartograph machine developed and formalized the mysterious technique or gaze of stereoscopy.[52]

Only once certain settings and adjustments were completed could the operator look into the stereoscope, and would the images begin to meld. The eye-piece pointers would 'appear as a single pointer, floating as it were in the field, while the two photographs are also combined by the eyes to form a single view giving the effect of relief' (Hotine 1927: 59). Hills and valleys suddenly took shape. Man-made features of fences and possible military features were just as suddenly revealed. Stereoscopy enabled a more realistic view of the ground below whilst it presented one that could not be normally obtained, even from the biological 'sense-organ' of the eyes. Photogrammetry was at once close to the conditions of 'natural observation' as was possible, given that all 'objects under distinct vision in nature' were understood to 'possess a perfect stereoscopic character' (Hotine 1927: 41).

But it was also significantly augmented. The aeroplane's mobility meant that what was once impossible to do – 'separate the eyes to a greater distance apart' – was now possible by taking two images at any required separation, limited only by the pilot's skill.[53]

Fusing these images together was quite a difficult process, particularly without the aid of any kind of technological assistance: 'It would be as wise for the average man to go through life with a patch over one eye as for the surveyor or soldier to attempt to deal with air photographs without a stereoscope' (H. J. Winterbotham in Hotine 1927: 2). Stereoscopic experts felt unable to understand the precise nature of the 'physiological or psychological process'. What they were certain of, however, was that stereoscopy could not rely upon an unruly body 'with no concrete anatomical machinery for effecting the purpose', and with 'the seat of relief perception' most probably lying 'entirely in the brain'; one needed to convert an impression into perception (Hotine 1927: 5).

Instructions pronounced that '[e]ach magnifying glass is placed by trial at a suitable distance to give a sharp magnified image of the photograph'. The separation of the two glasses was then arranged so that the eyes were over the inner marginal portions of the lenses. In this position the effective portion of the (convex) lens was equivalent to a formed prism, which introduced deviation in exactly the same manner as described above. The glasses could be held, one in each hand, or supported on a stand. To bring different areas of the image into perspective, the prints could be moved (Hotine 1927: 58). However expert the observer had become, '[t]he above directions should be followed rigidly' (1927: 54).

Those able to do this manually required considerable powers of 'convergence and divergence' as well as focusing power, the ability to call 'forth relief effects in hallucinogenic exaggeration from flat images' (Saint-Amour 2003: 356). Such powers were believed to depend upon bodily vigour and were 'naturally more powerful when the vitality is great'. They were skills and capacities not possessed evenly across the population.[54] For those without these capabilities, the stereoscopic mask could be utilized in order to fix one's gaze. The observer was to slide the prints around until the eyes fused them with much more ease than with the manual technique. The flexibility of the images meant that considerable eye strain would be experienced whilst fusion was being sought, and thus it was recommended that the desired distance optimal for fusion was achieved before looking into the eye-pieces.[55] Although quite strict boundaries of procedure were imposed onto the process, 'the fact remains that the human eyes do not obey any hard-and-fast geometrical laws and will, if at all possible, adapt themselves in order to tolerate any small departure from the conditions required by such laws'.[56]

Requiring mental, physical and technological expansion to produce or assemble the three-dimensional image, the stereoscopic aerial photo might seem distanced from airport securitization. Yet the biopolitical and biometric capture in examples such as US-VISIT is recognizably three-dimensional in its assembly of multiple sorts of data or information about the passenger. Facial image and characteristics are paired with fingerprint detail: 'Abstracted pieces of a person are taken apart, drawn into association, and displayed together' (Amoore and Hall 2009: 456). Like the three-dimensional models often made by British intelligence during the Second World War, the biometric record is the product of an 'assembly into one easily legible controlled and accurate three-dimensional statement of varied information'.[57] Both employ what Paul K. Saint-Amour (2003) describes as a 'virtual colossus', becoming tangible, graspable. The first to practise aerial photogrammetry found that the images they saw gave the effect almost as if 'suspended a few hundred feet above the tree tops'. But it was more than that. The images offered more than simply an improved fix on reality. Seeing an area 'as if he were examining a clay model', the subject becomes plastic, viewable from multiple angles and manipulable to action.[58]

Having said that, new forms of backscatter and millimetre-wave technologies re-envision the body-as-territory in another kind of projected reconstruction, not simply as 'multiple *perspectives* on a single whole', but 'precisely *projections* of something that would not otherwise be seen' (Amoore and Hall 2009: 455). Like the stereoscoped photos, these techniques permit a three-dimensional 'abstraction and a re-composition of the body's dimensions and densities', so making the object both 'knowable and amenable to intervention' (Amoore and Hall 2009: 456). Full-body scanners of X-ray backscatter technologies are being rolled out across many international airports – especially in the United States – because of their capacity to see and display information about the body beneath the clothing of passengers. In their 'virtualized, screened and projected reconstructions of bodies' (Amoore and Hall 2009), backscatter and millimetre-wave technologies do what the brain and experienced stereoscope interpreter sought to accomplish by envisioning something that would be otherwise invisible. As before, these techniques cut below the body's exterior. A 'virtual' or 'electronic' 'strip search', as the Electronic Privacy Information Center (EPIC) and the American Civil Liberties Union (ACLU) put it, backscatter and millimetre-wave technologies reconstruct apparently highly realistic images of the body beneath clothing, thereby revealing hidden contraband, explosives, drugs or weaponry. The technology can visually produce an image – like the X-ray pictures of luggage – that makes our outer skins invisible.

On the grounds of privacy and the right to dignity, EPIC, the ACLU and many other privacy foundations have criticized the TSA's plans, drawing on the technology's potential to degrade one's ability to shield one's 'unclothed figure from view of strangers, and particularly strangers of the opposite sex'. 'Impelled by the right to 'elementary self-respect and personal dignity',[59] the Privacy Coalition's national campaign (Privacy Coalition 2009) against 'whole body imaging' sent a letter to the TSA urging the stop of digital 'strip searches'. Harnessing the law of privacy according to a federal judge in 1976, their argument rested upon one's right to 'regard for his own dignity; his resistance to humiliation and embarrassment; his privilege against unwanted exposure of his nude body and bodily functions'.[60]

As Amoore and Hall (2009) argue, the use of the technology reveals an increasing trend to distanciate the border (Salter 2006), wherein pat-down searches and the touch of the security screener are deferred to a more distanced gaze, the passengers' reconstructed image is displayed and analysed in another room nearby. The obsolescence of the wand and pat-down search dissects the checkpoint function and experience, divorcing not only decisions from the border, as we have seen in the 'remote checkpoint', but decision and scrutiny from the body.

Conclusion

Today's modern soldier and pilot increasingly fight within a domain which is becoming known as the electronic battlefield, a battlefield supportive of 'total information awareness'. The aircraft and surveillance of cross-border mobilities initially sourced their information from one point, from above in the aircraft's perspective, or from a figurally similar point of the body in the case of airport surveillance. However, as stereoscopic viewing and military simulacra have combined, it is clear that the omnipresent view from nowhere and everywhere is becoming more and more possible. Untethered from the aircraft camera or the biometric fingerprint reader, for the 'full spectrum warrior' information is sourced from all angles – perhaps a drone aircraft on reconnaissance flown from over a thousand miles away, maybe a satellite image. Along with multilayered forms of three-dimensional digital dissection, the management of mobile bodies becomes holographic and projective, a convergence of information.

The projective holograms of the survey and the biometric image are attaining more form than we see here. With regard to the 'bodily functions' the Privacy Coalition wanted to hide, other forms of the aerial view address aerial life in a manner that does not simply capture or contemplate, but measures, judges and shapes, all in the blink of an eye.

Chapter Five

Profiling Machines

Introduction

In the human operator, writes Cambridge scientist J. F. Craik, 'at a particular instant (i.e. about 0.3 sec. after the end of the preceding corrective movement), a corrective movement having a predetermined time-course (usually occupying about 0.3 sec.) is triggered off' (1947: 57). The evidence upon which Craik bases this observation are studies of reaction time or the 'internal time-lag' of the human pilot-operator, which depends upon the 'time taken by the sense-organ to respond, for the nerve-impulses to traverse the central nervous system, for the appropriate response to be "selected" and for the nerve impulses reaching it to traverse the motor nerves' (1947: 57). This lag was found to be about 0.2–0.3 sec. Understood as a machine, the human operator, Craik concludes, 'behaves basically as an intermittent correction servo' (1947: 57).

The human operator as 'intermittent correction servo'? What a reduction and metamorphosis. It is to understand the human in the manner that one might observe an electrical circuit. The pilot-operator can respond to the 'feedback' of the machine with small correctional movements which occur in split-second time. Mechanical problems are overcome by drawing on the organic body, just as physical electrical motors and their circuits help make sense of the human's flesh and blood.

Prefiguring Norbert Wiener's famous work on person-operator 'feedback' and the birth of cybernetics, Craik and colleagues' work at Cambridge during the Second World War for the Flying Personnel Research Committee of the RAF translated the human pilot into circuits, gates and relays. A human–machine equivalence surfaced in the 'Cambridge Cockpit' – the Craik-designed machine which simulated the processes and manoeuvres of actual flight. The human operator could be easily modelled with a short chain of 'electronic components' consisting of a 'sensory device', a 'computing system', 'an amplifier' and 'mechanical linkages' (Hayward 2001). Each movement of the pilot and his capacity to respond to changes in the machine's progress could be compared to a synthetic operation. Like taking the car in for a tune-up, the aerial body was weighed against a machine: 'As the mechanic reports on the engine so must the doctor be prepared to assess the powers of the human machine and specify the work for which it is suitable' (*Lancet* 1918: 411). The object of this research was to develop a new science of *human factors* (Meister 1999) that would merge psychological, physiological and engineering expertise into improving the aerial body, with the hope that a leaner, more streamlined and reactive human could result. The British scientists aimed to create a subject capable of almost super-human feats of endurance, vigilance and enhanced visual acuity. In both Britain and the United States, the 'body would become machinized, or rather systematized – visible only in its reactions with and against a technologically mediated environment' (Hookway 2006: 37). Decision making, in other words, was to be subsumed and distributed through the body, and developed by training into ballistic reactions and automatic decisions formed pre-consciously.

Fast forward now to 2006 and a completely different context: that of a court case. An appeal was made in the British courts by the appellants on the grounds that they had been unlawfully stopped and searched by police whilst on their way to a peace rally. The police constables had used section 44 powers under the 2000 Terrorism Act in order to stop and detain the appellants for around 20 minutes. In the judges' summing up, an earlier case of racial discrimination performed by British airport immigration officers at Prague Airport (a case we discussed in Chapter Three) informed their decision. In order to question whether the stops had been unlawful (O'Malley 2006), whether the exercise of 'stop and search' was discriminatory, the judges drew upon the findings of the Prague case. The court found that if the stop and searches had been conducted because the appellants were Asian, then that would not have been a 'legitimate reason for its exercise'. Whilst they would accept that a person's ethnic origin was part of his 'profile', it was something else '(and plainly unacceptable) to profile someone solely by reference to his ethnicity'. Police would need to 'have regard to other factors too'.[1]

The appeal was dismissed partly on the grounds that the officers' intervention was not racially motivated. The search was legitimate because it drew upon an accepted practice of section 44 powers in airports, ports and high-security places where stops are routinely utilized. Transposing the context of street and airport, the *decision* for a 'stop and search' was all-important; it needed to 'be based on more than the mere fact of a person's racial or ethnic origin if it is to be used properly and effectively'.[2] The kind of decision this implied was not necessarily a slow, deliberate process or a complex system of targeting. A 'hunch', as opposed to 'something that can be precisely articulated or identified', was viable. The judges did give ground to the fact that age, behaviour and general appearance were considered suitable variables in the indication 'that a particular person might possibly have in his possession an article of a kind which could be used in connection with terrorism. An appearance which suggests that the person is of Asian origin may attract the constable's attention in the first place'.[3] However, the court left room for the possibility of an intuitive judgement – for the purposes of 'expediency' (see Pugilise 2006). This judgement would be 'undertaken, perhaps on the spur of the moment, otherwise the opportunity will be lost'.[4] The reason does not have to be based on grounds for suspicion, but it may be based, as Lord Brown had said, 'on no more than a professional's intuition'.[5]

The case at Prague's airport sets the legal precedent, imprinting itself upon the 'stop and search' case. When using their authority to select a person for further scrutiny – for 'stop and search' – the 'gut' reactions, instincts or intuitions of the airport police, immigration and security staff stand as a legally legitimate form of decision making. Whilst statistical and actuarial data appear objective measures to determine guilt or at least further investigation, as Ericson explains, they are frequently grounded on a 'basis of knowledge that is intuitive, emotional, aesthetic, moral and speculative' (Ericson 2006: 348; see also O'Malley 2006).

What we see here is similar kind of decision-reaction to that studied by Craik. Craik's adapting automatically acting pilot as servo-mechanism is reproduced in the police constable following the pattern of airport security officers. Their reaction is not necessarily thought through with conscious deliberation by slowly processing complex pieces of information; indeed, it is very difficult to explain or represent. It is reactive, ballistic; it is decision making on the fly. Quick judgements are made which are based on experience or on-the-job training. Slow cognitive judgements in both contexts are eschewed in favour of something different, something felt at quite different sorts of register, agitating and urging a decision to surface.

In the previous chapter, we examined an aerial gaze that sought to know, analyse and know more. Paralleling this view, however, is the tendency for the aerial subject to see and be seen with not much more or any less clarity than before, but by an apprehension of what they can do, and by techniques that accord priority to shrinking the distance between observation, decision and action – collapsed to the blink of an eye (Bishop and Phillips 2002a; Gladwell 2005). This chapter will explore how aeromobile populations are increasingly addressed through 'profiles' which, while patently calculative and quantitative, are able to pierce bodies at the level of the flesh, the embodied registers of stimulus, affects, feelings and emotions – the vital capacities of life (Rose 2006). This chapter is about the location of these capacities by the subsequent efforts to target them: how to determine, train or augment the aerial subject's capacity to decide, act and feel? How to prevent the tired and fatigued body? How to encourage attention? How to turn up mood to the required levels? The answer, as geographer Derek McCormack describes in quite a different context, is an attunement to a 'less immediately fleshy, and less tangibly human' sort of subject (2008: 426).

Imagining the Pilot/Passenger

Examples of Craik's mechanistic models are abundant in the first imaginings of the aerial body, specifically that of the pilot. David Pascoe's (2003) recovery of the airman belonging to a 'new race' shows how the differences between humans and machines became narrowed. In the development of pilot training and the emergence of the field of aerial medicine, the pilot was matched up with the machine, becoming one. The pilot increasingly became conceived as a 'controlling and coordinating mechanism', echoing the artistic appreciation of the aircraft's conduction of pilot agency through its tangle of gears and machinery. Robert Wohl describes futurist Gabriele D'Annunzio's ideas about the pilot imagined as 'technological superman whose bones and flesh have been transmuted like the wood of the propeller into "an aerial force" whose mission was to triumph over nature' (1994: 122). A body 'shaped and honed, strengthened and enjoyed, through activity; a body at one with a machine, swift and efficient' (Pascoe 2003: 127). And, just like the aircraft machine, 'this controlling mechanism must be in every way suitable and in the best of order,' stated the 'Medical Requirements for Air Navigation'. 'It is for this reason', the requirements continued, 'that only fit healthy human mechanism must be placed in charge of flying machines' (*Lancet* 1920a: 841). The pilot, in other words, became a mass of pipes, cables, pumps, propellers and struts, which had to be working in the most efficient way possible (Schnapp 1994). For Pascoe, Fokker's aircraft designs harmonized pilot, eye and gunsight; the first time in the air that the pilot was well and truly 'wedded to his machine; his gun coincided with his eye' (Pascoe 2001: 21). The successful flying officer would be one 'possessed of a pulse which gives the impression of a smooth, beautiful piece of mechanism' (*Lancet* 1920a: 838). Therefore an understanding was needed of how the 'human machine', which only 'operated properly at the lower levels', could perform at altitude.[6]

McCormack's investigation of the 1897 Andrée expedition to the North Pole, reveals the emerging science of the body in flight and in extreme environments; a body coming to terms with new technical fields of affect which re-engineered particular kinds of sensing bodies through 'pipes and cables, guide-ropes and drag-lines, and assorted technical devices' (2008: 426). Specific atmospheres and experiences of buoyancy, pressure and attitude would emerge from these encounters. The less tangibly aerial subject produced through such Spinozan technicities was premised upon classical imaginaries of the human as machine – a gear-box of intensities, feelings, circulations and flows.

Human machines, energy and fatigue

Opening up the pilot body could help one to understand the inner mechanisms and circulations that would translate into a more efficient pilot. Medical research was focused on the anatomy of the 'inner ear' – or labyrinth – and it was found that 'the seemingly bird-like qualities of the pilot' depended on tiny nerves (Birley 1920: 1252). In a not altogether successful comparison of bird and pilot, the labyrinth was seen as giving the pilot his appreciation of space and their movements within it. Along with the eyes, 'the great distance-receptors', RAF medical research officer James Birley positioned the labyrinth as a complementary visual apparatus. The labryinth produces an impulse that hardly passes 'the threshold of consciousness'. Such drives and stimuli 'are merely a wave in the great sea of impressions which together make up the state of consciousness'. Given the pilot's reliance upon the influence of visual sensation and accurate adjustment, 'the labyrinth gains its true significance' (1920: 1253). Through the science of the (un)conscious body in flight, the field of aviation medicine developed (Gibson and Harrison 2005).

The ability correctly to regulate and control one's nervous, respiratory and vestibular systems or Eustachian tubes was believed to be vital for an effective airman. He should be able to control his sense of distress, feeling of vertigo, the need to vomit and other forced movements. The admission of pilots to the RAF would be based on these sorts of capacities. Along with the circulatory pipes and tubes that made up the mechanisms of the aerial body, the pilot, the guns and machinery he controlled found their conceptualization 'as servomechanisms within a single system'. Indeed, the alleged father of cybernetics, Norbert Wiener, realized that because the mechanical elements of a machine could be understood with far more clarity than the pilot's psychology, 'we chose to try to find a mechanical analogue of the gun pointer and the airplane pilot' (cited in Galison 1994: 28). Servomechanical theory, Galison summarizes, 'would become the measure of man' (1994: 240). In Britain, Craik's prosthetic machines had to work at the speed *of* and *as* the power of thought, translating processes into words, numbers or other symbols; fulfilling a logic of reasoning, deduction, inference, and retranslating these symbols into external processes of action (Craik 1943).

'Mechanical Brains, Working in Metal Boxes, Computing Devices and Bombs with Inhuman Accuracy', said *Life Magazine*'s study of the Sperry Corporation's gyroscope-driven gun turret in 1944. The new turrets were represented by modernist painter Alfred Crimi, who had previously drawn medical frescoes of surgery and anaesthetics (Figures 5.1 and 5.2). His images presented the aviator as transparent – 'as though seen through an x-ray' – in a manner not unlike the backscatter visualizations discussed in

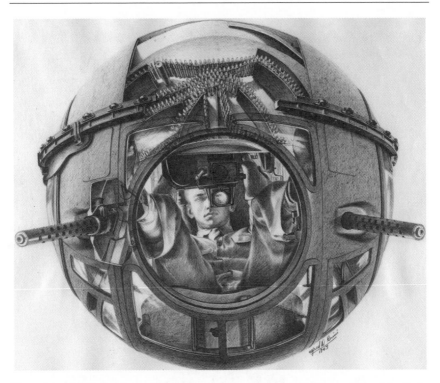

Figure 5.1   Crimi's illustration of the Sperry gyro turret. (Reproduced with permission of the Hagley Museum and Library.)

the previous chapter. Through the translucency one sees the pilot's control of the aeroplane. The 'human operator is surrounded by the machine, is intimate with the machine, becomes the machine, reining it in' (Mindell 2002: 63).

Fatigue was perhaps the greatest problem which faced the military scientists (see also Bissell 2008 on this relation), a state that they sought to diminish in stressed and tired pilots. High-level mental activities such as 'analysis of incoming impulses and the formation of correct responses' were believed to fail without adequate sleep.[7] For as pilots became fatigued they began acting more like machines whose energy – an equivalent commensurable with the body's fatigue – could waste away (Rabinbach 1990). According to a study by G. C. Drew, Craik's Cambridge colleague, when a pilot was fresh he was able to make considered judgements and regard the interconnecting circuits and instruments and how they might respond to his corresponding movements: '[H]e regards the task [...] as essentially a unified one, complex and having many clues on which he must base his movements, but still essentially unified.'[8] Flying 'by instruments' meant the pilot could

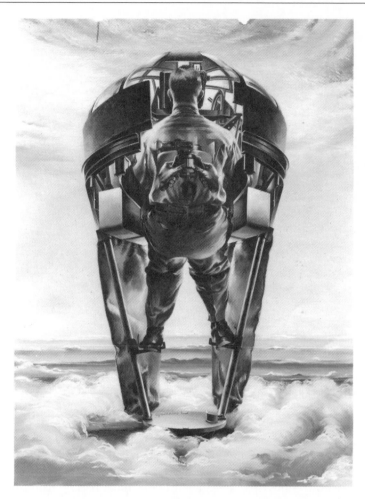

Figure 5.2   Crimi sees through the gunner's body. (Reproduced with permission of the Hagley Museum and Library.)

accept any movement of an instrument as a 'system of the behaviour of the aeroplane, and not merely of that particular instrument'.[9]

The fatigued pilot, rather than demonstrating an awareness of this system, 'tends to forget that he is flying an aeroplane, and regards his task in more simple terms as that of keeping six instruments to certain definite readings'.[10] Splitting up the task before him into the job of monitoring instruments, the pilot acts on a reduced 'stimulus–response basis, to movements of 6 discrete needles'. The consequence of all this is that the needles are regarded no longer as linked to the behaviour of the machine or even to the pilot, 'but merely as a stimulus demanding an immediate response from him to get the needle back where it should be'.[11] Strain and nerves were even variables of risk for air photo experts interpreting the images given to them: '[I]f the instructor suspects that the observer is nervous or flurried, he should leave him,' stated Hotine's influential manual. Stereoscopic observation was believed to require a 'mental calm at the best of times' (Hotine 1927: 69–70).

As an outcome of fatigue, the pilot is transfigured into the behaviour of a machine. This was one way of seeing the problem. And solving it, achieving resilience to the strains of flying and the onset of fatigue, drew upon familiar metaphors of the human–machine analogue. 'The nervous system, as a result of work or stress,' according to a report's authors, 'becomes depleted of its energy and power much in the way that an electric battery runs down through a drain on its circuit.'[12] Rest and appropriate treatment could enable the 'recharging' of these circuits. Little more than a 'human organism' led by its 'nervous system', the pilot could be 'refreshed and fortified by means of rest, sound sleep, appropriate diet and change of environment'.[13] Adequate rest and relaxation could provide enough 'resistance' to nervous breakdown when exposed to danger or stress.[14]

From the science of the pilot to the analysis of airport circulation, the scientific management of the aerial subject draws aeromobile populations into sympathy with mechanism and process, seeking to reduce effort and wastage in order to increase speed and efficiency. Consider how the passenger has been referred to by the aviation industry by the three letters PAX. According to Tim Cresswell, the PAX is essentially the 'transformation of mobile bodies into a legible record' (2006a: 238–9). The PAX is an abstraction, a simplification of real passengers, a generalization of what real passengers look like, think, feel or act. For one author, the PAX resembles 'a unit regarded as being of a basic standard, usually miniscule in size, somewhat lacking in both intelligence and general ability to find his way about (especially if he is a holiday traveller on a package tour)' (Donne 1988). As little more than a 'particle' in Virilio's terms (2005), such conceptions of the passenger draw upon a logistical science of operations and circulation management which position the passenger, as with the pilot experiments, as equivalent to a floating variable within a complex system of movement and processing.

In a commercial context, the Operational Research branch of British European Airways (BEA) in the 1950s developed 'formalized and integrative knowledges of sequence' (Mirowski 2001) that counted and calculated in order to develop an optimal sequence of events. Imposed upon real passengers and real people, translated into practice by refining future plans and restructuring the everyday processes and procedures of the airline to make it more streamlined and efficient, Operational Research was generative of a particularly simplistic and calculative passenger-body and airline-staff whose signatures were relative levels of productivity (Mirowski 2001: 178). In airports one such method for interrogating and refining passenger mobility was the use of card-punching techniques, pioneered by John Braaksma (1976) in the United States. This technique was also closely tied to a form of management called critical path analysis (CPA), which was used to better understand and optimize passenger as well as staff mobilities (Battersby 1964; Braaksma and Shortreed 1971; Lockyer 1969; Sealy 1957). The passenger and personnel experiences were subsumed into the representation of the line and the calculative variables of time and movement.

PAX, impulse and intent

Even though earlier models of airport behaviour worked on reducing the passenger body to the lines and pathways they would move through, newer forms of airport management, security and design are turning back to the body and its relationship to the impulses and motivations of the airport environment. In order to manipulate the passengers' desires, airport operators understand that different kinds of passenger will react differently within different situations. They do this by building up psychological profiles of potential customers. Some analysts describe this as 'tapping into the mindset' of the consumer (Behnke 2005), employing researchers to provide detailed statistical information on airport behaviour patterns – to build consumer profiles on the kinds of people who inhabit airports. Some academics have been working with airport authorities to help them understand the relationship between motivations and buying – a political economy of airport affect. Using data trails from airport management and operational support system software, information collected by airline reservation systems, as well as consumer spending habits from the airport outlets and shops, along with new kinds of airport baggage tags, tickets and gate passes, airport designers and managers may imagine, track and predict passenger behaviour within the terminal space (Albrecht and McIntyre 2006; Dodge and Kitchin 2004).

These possibilities or passenger potentials are mirrored in the construction of security profiling narratives, which scholars such as Mike Curry (2004) argue rely on the imagination of future stories. This means that when a passenger is 'faced with a particular set of circumstances', the system may predict how they 'would likely act in a particular way' (Curry 2004: 489). Each profile of passenger may be less or more likely to experience a variety of 'feeling states' (Newman et al. 1994: 11), which will affect their disposition to spend or not. As with the pilots, fatigue is a familiar problem that faces designers of airport consumer spaces, for if the wrong kind of 'affect-state' is maintained, the likelihood that passengers 'will spend, spend, spend' (Bates 2000: 45) is severely diminished.

In the security context, the process of 'tapping into the mindset' is becoming integral to the formal techniques of airport security. Racial and behavioural profiling techniques have emerged from Israeli and US security firms exporting their wares. These are premised upon formulations of the passenger-body-subject as not only susceptible but becoming-*dangerous* – an object within the circulatory system of air travel that threatens to undo it from the inside (Dillon 2003). Like the figure of an operational PAX, the security profile commonly stands in for the passenger. Originating in the realms of consumer marketing, the police and drug enforcement agencies, and later perfected by Israeli Airline El Al, passenger profiling has taken many forms and guises (Elmer 2004). Imaginations in the media have highlighted the problems faced by one who 'Flies while Arab', as ethnicity, race and nationality are feared as markers of judgement to indicate a terrorist or a threat. The killing of the Brazilian electrician Jean Charles de Menezes by police on the London underground in July 2005, who mistook him for a terror suspect, is a pertinent example (Pugilese 2006). Like the consumer imaginary, a profile constructs what Mike Curry has understood in terms of 'a narrative'. The security profile is more than simply an imagination of a subject with certain capacities; it also establishes 'the plausibility of that narrative' (Curry 2004: 488).

In this rendering the passenger is now imagined with long biographical histories which act as evidence of potential behavioural and personal indicators (Bennett 2006; Lyon 2007). By placing people into a specific category or a 'risk pool', one may then use this profile to predict – using known data to theorize the unknown. However, the security profile is now merging with the consumerist political economy expertise of the airport terminal. Passenger 'felt experiences' are being made both measurable, quantifiable (Adey 2008b; Lisle 2003) and increasingly suggestible of danger. In this trend, subjects are 'accounted for on the basis of behavioural potentialities, rather than on the basis of how they have actually acted' (Bell 2006: 160).

Airport security is becoming increasingly attuned to the feelings of passengers in new imaginations of the passenger security profile as emotional and affective. In these new contexts (Hawley 2007), decision making can be made without much deliberation; the passenger body, like the pilots, becomes a surface from which clues may be read. The inverse of Georg Simmel's (1950) appraisal of the face, in the latest modes of airport securitization, the face becomes the repository of the future or a reservoir of feelings displayed as they happen, anticipating futures which might be, or feelings they may want to keep hidden. The model for these style of address is no longer mechanistic, but biological. Charles Darwin's early investigations of primates discovered what he believed were universal dispositions to express feelings through specific facial movements, which he found in both animals and humans (Darwin 1892; Ekman 2006). Various followers and academics inspired by Darwin's work later found a number of specific facial movements expressed according to several different affect-states. Developments in psychology and neuro-psychology led by academics such as Silvan Tomkins (Sedgwick and Frank 2002; Tomkins and Demos 1995) and his student Paul Ekman have argued that feelings are actually constituted through facial mobility itself (Ekman 2001, 2003). In the new airport security context, these expressions are supposed to betray the movement of what is becoming understood as terrorist or 'hostile intent'.

Sorting

The aerial body and what it does have become open books to be scrutinized and read. Having established different kinds of passenger/pilot/terrorist, these imaginings are used as tools with which to sort, order, classify and exclude or include. But the question must be posed '*how to see?*' How to determine 'the right and wrong kinds of' passenger, the right and wrong sort of qualities?

Models that the aerial body must match are both old and commonplace today. The US Army's anthropometric studies illustrate the kind of standards which all aviators and forces personnel should match. For example, Total Arm Reach (TAR) is evaluated in the pilots and co-pilots to ascertain whether they can reach all switches and flight control, operating them through their full motions (Wilds 2006). Total Leg Length (TLL) is measured in the same way for foot controls. And a sitting height in excess of 95 cm is necessary in order to determine whether aviators can sit safely in each aircraft while reaching the flight controls in that 'normal sitting position'. Sorting out unsuitable aviators from others, the standards ensure that pilots are capable of accomplishing the tasks that maintain the cockpit as a 'powerful battlespace' (Wilds 2006).

Disposition

The sorting of aerial bodies is operationalized at a deeper level than the anthropometrics of the human body, but more in terms of their capabilities and capacities to act. '[W]hat type of officer is it desired to have in the RAF. We want to know what the finished article is to be before we can determine how it can be manufactured,' stated the first RAF committee of 1921, whose task it was to identify the characteristics and training of an ideal officer.[15] As far as possible, initial scrutiny panels could prevent 'the admission of:– i) the candidate who eventually will not care about going into the air, and ii) The Candidate who, in the event of war, will prove unsuitable for fighting service in the air'.[16]

Having established an understanding of the necessary tasks a mechanically understood pilot-subject would do, pilot selection worked backwards in order to define 'a certain minimum of capability of body, mind and character', without which 'no candidate should be admitted'.[17] Along with the 'mental and physical qualifications', the RAF committee prescribed the 'quality of character necessary for an Air Force officer as for every other officer in the service of the King'. Having something to do with the relatively fuzzy notion of 'moral temper', *character* was pinned down as 'the quality of a gentleman'. Gentlemanly conduct was of far more importance than any 'degree of wealth', 'social position' – a 'certain character' was necessary. The future air wars created a potential space in which the right kind of aerial body, with the right kind of qualities, was crucial (Wells 1995).

Pilots were commonly conceived as 'aces', characterized by their '"reckless courage, fool hardiness, contempt for death"; a breed whose great exploits – flying across the water, over the desert, against the odds – exist in inverse proportion to the fragile contingencies of their machines' (Pascoe 2001: 23; see also *Lancet* 1918). '[E]ndowed with fearlessness, a love of adventure and sport, dogged determination to overcome difficulties, and, perhaps most important, except when fatigued, a love of the air' – all of these qualities proved elusive to distinguish. Such 'temperamental fitness' was seen as 'peculiarly difficult to assess'; even a medical exam 'affords but little guide in this respect'. The RAF committee noticed that there was 'something more vital in the human element that can only be detected by certain persons who are themselves natural and experienced flying instructors'. A special sense had to be nurtured, a sense 'akin to that of a Water Diviner [...] it is an art that can be fostered and imparted'.[18]

Therefore by 'what method is the choice to be made?' the original RAF committee had to ask. How to give 'hope of getting the right type of man, physically, mentally and morally'? Prescribing a 'somewhat complex method of choice', the Cecil Committee suggested an 'enquiry into the physical and moral qualifications of candidates' which would be combined with intellectual tests, amalgamating 'processes of selection and examination'.[19] The most thorough of these was the medical exam developed by RAF medical officer Martin Flack (1919, 1920), who engineered a series of trials and tests for the recruits.

After an initial examination, a pilot would undergo an 'elaborate medical inspection', which was designed to do more than ascertain his fitness to be airborne; it also assessed his physical capabilities with a 'view to skill as an aviator and endurance under conditions of active service'. The person conducting the inspection would use what they could immediately see, and examine along with the medical history the results of the qualifying examination and a full statement of the marks. All these inputs would be presented to the Board of Selection. With a report, if obtainable, 'as to the moral, mental and physical qualifications of each candidate from the teachers under whose tuition he had been as a body', this would contain aspects of 'character temperament and aptitudes'.[20] In the United States, the standard nine or 'stanine' test would play a similar role in testing various physical and psychological attributes (Sherwood 1996, 1999).

The 'elaborate medical' inspection meant testing several critical faculties of the pilot which burrowed down into the minute depths of the body's physiology. Four reasonable degrees of visual acuity were required with the ability to judge distances. These included being able to demonstrate a normal refraction and a healthy condition of retina and intrinsic muscles. For the second test 'it is necessary that the aeronaut should possess good "binocular vision and normal balance of the extrinsic" muscles' (*Lancet* 1920b: 1026). In an eye examination, the 'intrinsic and extrinsic' muscles of the eye were subjected to tests of their lack of tone or over-action, affecting acuity as well as producing 'conflicting, delayed, or wrong interpretation of messages sent from the eye to the brain' (*Lancet* 1920b: 1026).

Performance tests uncovered these kinds of inadequacies. A patient was asked to fix a pencil in their gaze, and then cover one eye with a card before it was uncovered. If the eye moved on, a problem of balance was revealed: 'A perfectly balanced pair of eyes will remain fixed on the pencil whether one is covered or not, whereas movement inwards or outwards on uncovering shows lack of balance under convergence' (*Lancet* 1920b: 1026). By *doing*, and conforming or not to models of acceptably conscious and unconscious capacities to perform, the tests sought to 'comb' out individuals who should be admitted and those they should discard.

Balance tests were an idiosyncratic performance which could display the qualities evident for different kinds of air officer. An applicant was asked to balance for 20 seconds, 'an ordinary walking stick perpendicularly on each hand', and then with each stick separately, on the 'toes of each foot, always keeping the eyes fixed on the top of the stick'. If this could be completed successfully, the applicant could show the 'basic requirements for training as a bomber pilot'. Proceeding with the tests once more, but on this occurrence with a much smaller stick and for half the time, if the applicant could successfully find balance, he 'shows the basic requirements for training as a Fighter Pilot'. Indeed, if an applicant could master both, 'he should prove to be a good all-round pilot and possibly a valuable instructor'.

The study of pilot reaction time became a primary means of measuring and testing for the purposes of selecting and improving the aerial pilot-body. As Jonathan Crary writes on the context of James McKeen Cattell's influential research on the body's reaction times, the tests brought the pilot into a relation of *subjection* to a new world of tests and experiments. The pilots needed to meet and accommodate 'a generalized attentive expectancy and kinetic adaptation to machine speeds and rhythms that differed dramatically from those of the body' (Crary 1999: 309). These were worlds of violence, speed and extreme sensation, a whirlwind of conflicting impulses and afferent stimuli.

Different modes of piloting were believed to necessitate different sorts of 'temperamental constitution', and 'certain outstanding motor and cognitive (intellectual) characteristics'.[21] For instance, good fighter pilots should be 'be quick both physiological and mentally; ready to jump to practical conclusions

Table 5.1   Tabulation of the main temperamental qualities desirable in pilots

	Fighter	*Bomber*
Tempo	'Dashing', quick reacting type; Less intellectual	Slower, more cautious type; More intellectual
Perception	Seeing the whole situation	Analytical type
Thought	'Observational-objective'	'Abstract-systematic'

Source: D. J. Dawson, 'Pre-selection of flying personnel', Flying Personnel Research Committee, FPRC 48b, 1939, p. 1, AIR 57/4, The National Archives.

in advance of the actual evidence; able to sustain intense activity for relatively short periods; able to "get up steam" very rapidly (i.e. with a rather horst adaptation or warming up period)'. They should be 'strongly competitive; with a rather scattered type of attention'. Directed outwards towards the environment, they would need to 'keep step with rapid changes of stimulus', as well as 'changes in the rather narrower field presented from within the aeroplane itself'.[22] On the other hand, the good bomber pilot should be much more deliberate, 'more prone to act decisively for reasons or with a plan in mind rather than on risks'. The emphasis was on maintenance of performance for long periods of time, 'with a concentrated type of attention in which the field is relatively small and the changes often minute; especially resistant to the fatigue of prolonged activity'.[23] These tasks were identified by a profile tabulation of main temperamental qualities (Table 5.1).

Instructor reports indicated that around two-thirds of the rejections after initial training were attributable to a 'large group of undefined factors'. As F. C. Bartlett explained, 'A man can for instance carry out all the control movements required and can perform them efficiently, but under certain conditions he fails to produce them when they are required, or he does that badly.'[24] What is more, this brand of scrutiny and sorting needed to continue throughout the pilot's life in the air forces. Fatigue, already discussed as a persistent problem for aviation psychologists, could build cumulatively throughout a pilot's career. The task, again, was spotting the evidence for its onset, over which the pilot had little control, being unable either to hide it from colleagues or to maintain his own performance (Flack 1921). To an outside observer, the fatigued pilot could be pronounced 'stale or bored, indifferent'. As the condition increases, 'emotional symptoms appear; depression, or irritability; or outbreaks of irresponsibility and undisciplined action as an escape from his failure to carry on acceptably'.[25] Through tests, Flack could develop modelled dispositions of the likelihood of fatigue.

Once more, balancing became a form of diagnosis owing to its ability to target the 'subject's fine control of his muscular activity', which existed somewhere on a 'borderline between the voluntary and the involuntary'. For as the pattern of behaviour appeared to change with variations of 'physical and mental conditions such as concentration of attention, of muscular effort', it was argued that 'normal tremor might serve to measure and identify the onset of cumulative fatigue'.[26]

Pilot selection therefore relied upon much more than rigid and prescribed objective measurements and analysis, and involved the eye and scrutiny of an experienced professional. Careful not to suggest 'introspection' among pilots, it was put forward that pilots could record their own 'stress hours' for a medical officer to analyse and interpret. Of more real benefit were medical officers living with the pilots. They would 'rely upon their impression of the psychological state of pilots as they meet them in the mess', making it easy for the officers 'to know and observe all officer members of aircrews'.[27] Living in the mess was an 'obvious advantage'; the officer 'should be a good mixer', and 'should have been long enough with his squadron to know the flying personnel well', and thus able to tell whether a pilot was falling into an anxiety state: '[I]t short circuits the risk of suggesting ill health.'[28] The medical officer's observations would be developed through practice, exercise and, importantly, 'sympathy',[29] resembling the Cecil Committee's prescription for a 'a flying officer of great experience, who has shown himself to be a student of human nature, and a good judge of the type of officer required for the Royal Air Force'.[30]

On the other hand, attempts were made in 'boarding out' problematic recruits even before they were accepted for training, particularly those on grounds of 'mental dullness, personality maladjustment, neurotic or prepsychotic conditions, anxiety and inadaptability'. This would occur within a week of a recruit entering a Women's Auxiliary Air Force (WAAF) or RAF recruit centre. Creating a standardized system of investigation and comparison to psychological profiles was the suggested course of action.[31] A report concluded that questions on deciding the 'temperamental suitability for a particular trade or the general nervous stability of the airwoman is [...] left to the interviewer with no specialist training in this direction'.[32] It was thus advocated that specialist psychologists should assist, for whom the 'personal interview' was their primary method of detection.[33] Heavily influenced by industrial psychology on 'training wastage',[34] the Air Force sought to develop a thorough and organized series of filters.[35]

The tests were about more than how one could perform, or at what level of performance one could be relied upon to act. They were about disposition. Ground crew recruits showing signs of initial nervousness would be sent to psychiatrists for further tests. Nervousness, 'deliberately intended to be a very comprehensive category', included 'unduly shy individuals', 'stammerers', 'blushing'; 'fidgetiness'; those who were 'over-talkative' and 'apparently self-centred'; 'aggression'; 'people with a grudge'; 'mother's darlings'; 'anyone pleading illness; and 'those who seem preoccupied'.[36] A disposition even towards air sickness was explored due to the incapacitation of subjects at critical moments during flights.[37] What was one's disposition towards fatigue, towards sickness, towards poor performance, towards susceptibility?

Discovering these sort of embodied character traits – the physical and psychological dispositions necessary to perform one's role as airman or -woman – shares many of the characteristics witnessed today in airport security processes that attempt to sort out high- from low-risk threats. Akin to the evolution of pilot testing, the processes of capture have become increasingly bio-physiological in both their target and processes of observation.

Faces and facets

Since the eruption of aerial terrorism in the 1970s, a growing trend in airport security has been to harvest as much capta (Dodge and Kitchin 2005) as possible about potential and actual passengers, whether for the purposes of identifying drug smugglers, terrorists or illegal migrants (Feldman 2007: 333). Of course, the collation of information about frequent flyers both mirrors and parallels this process in schemes such as the now-failed Clear programme in the US. Indeed, archived litigation in the 1970s is now useful testimony to the use of increasing amounts of information shared between law enforcement agencies that could inform the denial of a passenger to a flight. As we have seen, however, the use of profiles and typological norms to filter out anomalous and threatening behaviour is a key condition of aerial security's efforts to secure its circulations, resonating with wider contemporary forms of security that attempt to 'sense rogue behaviour' in order to anticipate societal threats (Dillon and Reid 2009).

Debates over drug profiling in United States' courts tell us a great deal about the background of airport security. The airport 'drug courier profile' is an important case in point. First developed by the Drug Enforcement Administration (DEA), it is a way to inform 'probable cause' and the arrest of an individual or even the suspicion necessary to justify an investigative stop. The DEA profile did not have an easy route to legality, however. According to one court's judgment, probable cause was deemed inappropriate for two main reasons. Firstly, the integrity of the profile was far too 'amorphous' to be accepted as any sort of legal standard. The physicality of the profile was not even evident: '[T]his profile was not written down, nor was it made clear to agents exactly how many or what combination of the characteristics needed to be present in order to justify an investigative stop or an arrest.'[38] What's more, using a profile to determine probable course jettisoned the holistic information necessary for legality. The profile was portrayed as an 'individual layer of information' rather than the 'sum total and synthesis of information'. Probable cause in *United States* v. *Van Lewis* (1977) was rather 'the sum total of layers of information', it was a 'synthesis of what the police have heard, what they know, and what they observe as trained officers': 'We weigh not individual layers but the "laminated" total'.[39] Taking the layer rather than the whole could skew and distort the entire and complex picture of the facts. However, running alongside these examinations is the increasing use of less cognitive forms of judgement, and far less detailed and composite kinds of information gathering. These are based, like pilot selection, on a fleshy and biological imagining of the passenger-subject.

Knowledge of reactions, physiological indicators of 'offending acts and gestures' are pertinent here, and common across the wider governmental apparatus of security and infrastructural protection. In the context of Mick Dillon and Julian Reid's (2009) analysis of the Department of Homeland Security's (DHS) Common Operating Picture for critical infrastructure, understanding the corporeality of the border is key as the body enacts an 'auto-confession' testifying, 'along with our documents, about our intentions, character, utility, moral quality, and social and economic origins' (Salter 2006: 183), and is deciphered according to its 'deceptions' and 'abnormalities'. Running adjacent to the diagnostic analysis of 'archived guilt' (Feldman 2007), other forms of pre-emptive governance act without recourse to prior knowledge, histories, background trails or genealogical baggage, but the intuitive and physiological knowledge of capability we saw engendered in pilot selection (Thrift 2000a).

Addressed at the analysis of the 'divergence from a normal pattern of behaviour' (Dillon and Reid 2009: 143; Pugilise 2006; Weber 2007), expert in facial emotions Paul Ekman and his colleague psychologist Mark Frank have been providing pro bono work for the Transportation Security Administration (TSA) program known as Screening Passengers by Observation Techniques or (SPOT), all under the umbrella of DHS project 'Hostile Intent'. By drawing on Ekman's Facial Action Coding System (FACS) developed in the late 1970s, which catalogued hundreds of expressions of emotions, and by using something along the lines of the FACS Affect Interpretation Database (FACSAID), agents are trained to look for how feelings surface in minute and instantaneous gestures drawn on suspect's faces such as pursed lips, raised eyebrows and many more, especially those emotions people try to conceal (Adey 2009; Ekman 2001; Frank and Ekman 1997; Tsiamyrtzis et al. 2007).

Identifying and then sorting out apparently dangerous or 'high-risk' passengers for increased scrutiny or ejection because they display the wrong sort of qualities and characteristics draws heavily upon the discovery of intention. Intentions may be expressed unintentionally through feelings and then leaked away by what Ekman terms 'micro-expressions' (DePaulo et al. 2003; Ekman et al. 1999): those body movements that occur when one tries to hide an emotion. For instance, how 'a flash of anger' can quickly strike and pass from someone's face. These affects are named by the TSA as 'involuntary muscular responses'. According to proponents of this scheme, because the expressions are apparently involuntarily, they are incredibly difficult to defeat. In this manner, their techniques rely upon the contingencies of the passenger's future imprinting itself upon their present through the transition from intent into feeling (Dillon 2007a, 2007b).

In the wider context of policing and terrorism, such a process has become a common means of determining and selecting passengers for exclusion or further scrutiny. In his 2001 review of the operation of the UK Prevention of Terrorism (Temporary Provisions) Act 1989 and the Northern Ireland (Emergency Provisions) Act 1996, Mr John Rowe QC celebrated the 'intuitive stop' – the value of which was impossible to 'overstate' even when 'there is no intelligence or information upon the individual'. In this scenario, the police officer, 'with his or her training and experience, receives the impression on sight that the passenger may be of interest'.[40] The intuitive stop is 'made "cold" or "at random"', although Rowe found it more accurate to use the 'words "on intuition" – without advance knowledge about the person or vehicle being stopped'. Expert knowledge is required for these sorts of investigations, the reason for which 'cannot be explained to the layman'. According to Lord Carlisle's review, 'the trained and ingrained instinct of experienced officers still has real value. [...] Different officers operate with their own "feel" for what they see and hear.' The intuition described 'cannot be rationalised'.[41] As Salter notes, police officers and border agents 'must rely on "gut" feelings' (2008a: 10–11).

Much like the procedures of the pilot scientists who attempted to sort out candidates, these new forms of airport security effect a series of practices aimed at distinguishing one person from the next. This, too, is based on the *performance* of the passenger, who is placed under laboratory conditions. Salter's (2007b) play upon Foucault's notion of the confessionary complex explicates the interrogatory procedures of border control which entails the passenger to confess who they are and other aspects of themselves. For even as airports are compared to laboratories from which new technologies and procedures are tested (Fuller and Harley 2004), through verbal conversation and other forms of incitement, the passenger is poked, prodded and disturbed in order to reveal a sort of truth.

Through further interaction with trained detection officers and airport screeners, these forms of scrutiny play upon the officer's intersection within 'participatory thought' arising out of expressive-responsive bodily activities, as Thrift has discussed (2005b, 2006). A technology such as Suspect Detection Systems' *SDS 1000* (Brinn 2005) uses other kinds of monitoring by most notably asking for verbal responses to its questions. The questions posed usually encompass particular words that are intended to agitate the guilty respondents and activate certain bodily responses such as heart rate and temperature. SDS have developed a word library which they believe only terrorists will respond to. These include words that name specialized materials relevant to terrorist activities such as the making of a bomb.

The behavioural confession is 'capable of having effects on the subject himself' (Foucault cited in Salter 2006: 58).

Language becomes used both as a way to incite expressive feelings and as a way to capture them. The 'voice' of the passenger – which 'is the sign of pain and pleasure' (Agamben 1998: 7) – allows security offers to attune to the pre-linguistic undertow of language constituted in pitches, rhythms and amplitudes which may be coded into waveform patterns, or formed in further facial expressions and other kinds of body language. The discourse of truth is therefore solicited by incitement. Mobilizing what Jasbir Puar and Amit Rai (see also Anderson 2007; Massumi 2005b) usefully describe as 'the affects of uncertainty', 'the blurring between patriotism, heroism, betrayal, fear, cruelty, pain, and pleasure through the encasing and knitting together of modulated rhythms of the mediatized body of technoscience)' (Puar and Rai 2004: 90), behaviour detection incites or stimulates reactions. In suspects' effort to control themselves, the anticipation is leaked which makes the very future they were intending to avoid come true. The very effort to control those bodily signals compounds the emotions the passenger attempts to suppress, and thus, 'in a strange paradoxical feedback, activates the future at every turn' (Parisi and Goodman 2005). From this perspective, behavioural profiling works less by acting in advance and more by soliciting and, in a manner, 'letting go' (Foucault 2007); the 'hostile passenger' is allowed to reveal themselves, and the officer's feeling develops and surfaces. The behaviour detection programmes and techniques then do not simply allow the body to speak by reading it (Aas 2006), but rather they develop an apparatus or a machine, as Cote-Boucher puts it, 'that renders visible and generates speech' (2008: 145).

Modifying

The German experiments discussed in this book's introduction are positioned at the most extreme end of a continuum of tests on the aerial body which have evolved from the Second World War. Their example reveals the stripping away of rights and the values applied to the human – a body which has no value – who is *asocial*. But this stripping away is more suggestive of the submersion, literally in the *Versuchpersonen*'s case, of the aerial body into environments that hurt and threatened, stretched and improved their capacities of survival and behaviour. Unmistakably opened up in their autopsies to scrutiny and intervention, the Luftwaffe's stand-ins stand in for the aerial body who is produced, adapted and improved. The Pilot/Passenger becomes, in other words, a figure that can be trialled and tested, to be ethologically experimented with, improved upon and trained by the augmentation of their behaviour and the formation of repetitive and habitual 'initial reactions to movement in an unusual and disturbing environment'.[42]

Laboratory conditions, fields and force

The cockpit and the airport terminal are conceived and precluded as experimental domains. As propulsive spaces they are designed to pull on invisible pulleys, plug into the circuits, pipes and cables that make up human physiology or provoke and agitate eruptive animalistic potentials. The technical milieu of the scientist, the security official or the experiment forms what Georges Canguilhem would describe 'as a proper milieu for comportment and life, the milieu of man's sensory and technical values' (2008: 119). J. F. Craik even describes his Cambridge laboratory as a place extending his body into touch with the small and the microscopic as well as the far-away:

> [T]his small, dark room, cluttered with apparatus and a tangle of wires, is a place of infinite possibilities and far horizons, full of pathways leading me into the unknown. Its instruments extend my eyes and limbs – allow me to see living cells, the sound waves of my voice, the impulses in a living nerve its machines enable me to mould and carve metal as if it were putty. (Craik in Hayward 2001: 297)

Training and routing the body through augmented spaces and adjusted through various kinds of prosthetics, '[i]n both a physical and psychological sense', writes Sherwood, has lifted the person from the ground and 'transformed him into an "air-minded" individual' (1996: 38). For F. C. Bartlett, Craik's superior at Cambridge, the problem of '[w]hen a man moves with a machine' was the inevitable 'mass of proprioceptive impulses coming from changes of stress and tension at various points of the body surface, from underlying tissues and from other bodily sources' (1943: 250). When moving with and in machines or the environments discussed, Bartlett's simulated aircraft really felt like it was moving – a 'combined performance of the machine'. The issue was how to control and use these impulses, as well as how to survive them.

At Dachau and Cambridge, British and German military scientists created artificial spaces, simulating environments or conditions that would push the body to its limits. In Dachau, Dr Sigmund Rascher's cold-air experiment saw the building of a $2 \times 2 \times 2$ meter wooden tank which was lined with metal and filled with water (Poszos 2002; see Figure 5.3). Kept somewhere between 2.3 and 12 degrees Celsius, the subjects were dressed in flying attire, including a lifejacket. The subjects were naked in other studies. Rectal and skin temperatures could be recorded, and blood and urine diagnosis were also made. Methods to keep the subjects alive were employed after exposure to these conditions. These included the rapid re-warming of the subject by a hot bath, a heated sleeping bag, massage, packing in blankets, and, even more disturbingly, as Agamben (1998) notes, body-to-body

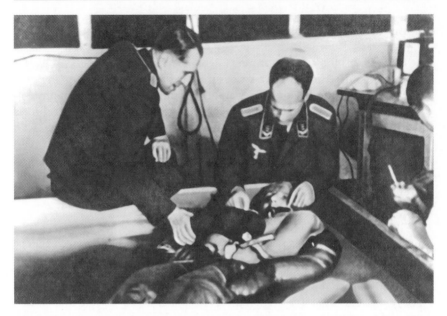

Figure 5.3   The Nazi hypothermia experiments at Dachau. (H.M. Hanauske-Abel, 'Not a slippery slope or sudden subversion: German medicine and National Socialism in 1933', *British Medical Journal* 313 (1996): 1453–63. Reproduced with permission of the BMJ Publishing Group Ltd.)

re-warming with some subjects placed between two naked females, resulting in forced sexual intercourse (see also Cohen 1949).

The problems of high altitude necessitated solutions that exposed the body to the experience of acute simulated forces and environmental conditions. Both German and Allied scientists used low-pressure environments that simulated the debilitating condition of 'air hunger'. High-altitude flights often meant pilots had to take two breaths in order to acquire the same amount of oxygen as they would with one breath on the ground.[43] Pilots would have to therefore work a lot harder in order to breathe sufficiently, a problem magnified given the other strains on them during flight. Lack of oxygen could be disorientating, 'dull perception' and add to the build-up of fatigue. Boeing's expertise in super-pressurized cabin spaces informed the conditioning of military pilots. Oxygen masks became the most popular method of overcoming the impact of the atmosphere upon the body's performance. Oxygen was piped to aircrew through face masks, forming a more minute bubble of safety that would allow the pilot to survive.

Gravity was one of the strongest and most investigated forces in terms of its effects on the acceleration of the body in flight.[44] Considering the aerial body in its subjection to the radically exterior force of gravitational acceleration, British and American scientists explained,

> A body at rest upon the surface of the earth is subjected to the pull of the earth's gravity proportional to its relative weight. The speed of a freely falling body adds to this normal pull by the development of an increased inertia. […] If the straight course of a falling body is altered, the exaggerated tendency to continue downward will persist as an 'acceleration' along the line of the direction of gravity. The strength of the force which develops will depend upon the inertia at the time of the change in direction and the abruptness with which the change occurs.[45]

The import of such fields was felt within. G-force accelerations reached the venous blood, throwing it into the lower part of the abdomen by 'engorgement of the systemic and portal systems'. Venous pressure in the cranium would be 'correspondingly reduced and there will be reduced filling of the auricles', whilst the 'increase in venous pressure in the lower part of the abdomen will impede the return flow from the legs and will cause a rise in venous pressure'.[46] Under the influence of these forces British and American scientists reasoned that 'arterial pressure should show changes resulting from the application of accelerations along the long axis of the body'. Thus, any accelerations that 'centrifuged blood into the lower parts of the body' could be determined by the measurement of arterial pressure.[47] The body was dissected by axes of force, acceleration and pressure.

The problems here are those of Peter Sloterdijk's (2009) 'being-in-the-air'. Experiments saw pilots stripped back to their skins and endowed with new ones. 'Suspended in an environment incapable of sustaining human life itself' (Wells 1995: 52), technological and affective prostheses tightly or loosely fitted – even internalized – saw the aerial body suspended within a new environment of improvement, from the pressure chamber to the second skin of the G-suit. For the rest of the chapter, we will see how pilots and passengers were endowed with specific spatio-atmospheric environments of the aircraft, the airport or the camp, not simply to keep them alive, but in order to manipulate, shape and ultimately improve their behaviour. As with the construction of super-pressurized spaces, or gravitational accelerations, by constructing an immersive or ambient environment, what John Allen (2006) describes as a 'particular atmosphere, a specific mood, a certain feeling', the scientist's production and simulation of an aerial milieu was able to render material and psycho-social affective fields. These were fields of 'potential attraction, collision, orbit, of potential centripetal and centrifugal movements' (Massumi 2003), whose gravitational forces and distortions were capable of shaping the body's capacities.

Pushes and pulls

In the late 1930s, pilot trainers in the RAF medical branch were approached by drug companies to try out particular chemicals which promised to improve an aircrew's stimulus reactions, alter their *attitude* to complex and difficult tasks, and, importantly, manipulate their emotional and affective state in order that they might face danger differently. Of endocrine-vitamin therapy, one company wrote to the Flying Personnel Research Committee proffering the apparent benefits of their new drug. These included: reducing nerve strain; imparting tranquillity and a 'spirit of calm confidence'; a 'maxima' of concentration and mental alertness; the allying of emotional instability; the 'capacity' to decide instantaneously at particular movements; confidence in the individual and his colleagues; a sense of uplift and wellbeing; and even practical immunity from infection because of 'increased resistance'. Even the awareness that by taking the drug one would gain superiority over one's adversary by being able to think a split second quicker was a bonus: 'That alone cannot fail to impress itself on the individual's personality.'[48]

Amphetamines, or 'Speed', such as Benzedrine, were tested as a useful method for 'greater fluency and quickening of mental processes, greater effort and persistence of effort and restoration of interest'; 'an appreciable reduction in subjective feelings of fatigue'; 'elation [...] wellbeing, cheerful-ness, freedom from doubts or anxieties, absence of tenseness and sense of relaxation'.[49] Whilst Bartlett's multiple investigations of Benzedrine showed that it didn't do quite what it was purported to, the drug had considerable 'behavioural and emotional – that is, strictly subjective – effects'. R. H. Winfield, who took further examinations of the drug with aircrew, saw that in order to reach 'peak efficiency', Benzedrine became a natural solution to help sustain pilots with 'the highest morale', which depended upon '"the sum total of all those things that affect the minds and bodies" of the avia-tors' (Rasmussen 2008: 65). 'Speed' was a quick route to the ideal modula-tion of morale, giving airmen 'the right combination of optimism and aggressiveness to reach "peak efficiency"' (Rasmussen 2008: 65).

Along with chemicals, prosthetic environments augmented pilot practice and behaviour. These were the flight trainers such as the Link trainer and the Cambridge Cockpit, which diagnosed the effects of fatigue upon the rage and irritation expressed at the trainer's 'feedback'. The stabilizing char-acter of the aeroplane was often interpreted as both 'friendly or hostile' or 'felt as a personality to be antagonized when the plane is forced into unusual manoeuvres'.[50] Our consciousness of will in another person, cyberneticist Wiener argued, is just that sense of encountering a self-maintaining mecha-nism aiding or opposing our actions. By providing such a self-stabilizing resistance, the aeroplane acts as if it had purpose, in short, as if it were inhabited by a gremlin (Galison 1994: 246). The machine's feedback, in other words, was understood by the fatigued pilot with intention, *projecting* onto the aircraft 'any errors of which he becomes conscious'.[51]

Like the breathed gaseous and aromatic environments which immersed other 'airspace users' (Sloterdijk 2009: 95) in the high-altitude experiments, contemporary airport design spatializes the emotional and affective stimulus of Benzedrine and other drugs in an immersive yet non-chemical ambience or atmosphere (see also Bissell 2008). By the 1960s, architectural critic Lionel Brett had ironically found the airport an anaesthetizing experience, one that acted as a 'sedative, bromide' or an 'antidote to feeling' (1962: 6).[52] Fatigued and dulled by the strains and processing of the airport, Brett called for drama and excitement – for airport design to stimulate sensation.

Different stages in today's airport processes are seen as 'hurdles' to accomplish that modulate emotions. The airport is envisaged as a landscape of differential mood and sentiment as many airports take account of what they see as a time-space-processing dependence on the distribution of feeling throughout the terminal space. Here we see an economy of affective potential distributed by design. Basic structural solutions are employed to shape and direct passenger mobility by providing architectural clues or triggers to the passenger in a way that does not require thought but rather forethought (see Thrift 2005b, 2006). Part of this solution has been to understand a dimension of the temporal extension of affect; the projection forwards or the 'feeling for potential'.

Given the relative states of anxiety associated with different spaces within the terminal and different processes, the propensity to spend money is a primary concern of airport designers, who attempt to create an 'atmosphere of point-of-sale' (Sloterdijk 2009: 94). Strategies such as these attempt to create a 'powerful' and 'persistent urge to shop for something immediately' (Omar and Kent 2001: 235). Terminal space is also populated by an array of wayfinding technologies, signs and symbols (Fuller 2002), which can require complex and sometimes very simple cognitive thought processes in order to make sense of the flight information displays, to figure out where the toilets are, how to get to the departure lounge. The affective forces of the terminal accelerate bodies with centrifugal energy outwardly in particular directions.

These new kinds of airport design initiate unconscious 'trigger' or 'pressure' points that even occur before cognitive thinking. Take, for instance, Richard Rogers' Madrid Barajas Airport Terminal Four, where signage and ambient colour schemes initiate unconscious triggers, pulling passengers through the building, or lending familiarity. Elsewhere, 'dual carriageways on airport approach, escalators, stairs, entrance and exits, check-in counters are (ideally) designed and located so that they provide a "natural signing" to their functions' (Fuller, 2002: 135–6). For Fuller, signage systems 'urge' us forwards; the arrow points, it gestures, it executes 'action plans' that somehow lie behind our thoughts or the doctrine of corporate wayfinding philosophy, where decisions are facilitated, executed and 'transformed into behaviour' (TCRP Report in Fuller 2002). Signage may be felt intuitively at an unconscious and affective register, experienced before reasoning through the body's attraction and repulsion.

In 2008 the Transportation Security Administration launched a new concept prototype of their evolution in airport checkpoint design and function. Tendered by IDEO, a self-titled 'human-centered' design practice with expertise in initiating organizational change, the TSA's new concept offered a central new benefit for passengers experiencing the airport security check. Security was now and significantly 'calmer'. Pre-screening Preparation Instruction Zones use 'architectural features, simple signage and instructional videos, and "ambassador" staff to create a calming atmosphere and more efficient throughput by instructing and directing passengers for efficient screening flow' (TSA 2006: 95). Far from independent of one another, a smooth flow of passenger mood can mean a smooth through-flow of their mobility.

Behaviour detection was intended to remove some of the more hassling and stressful features of direct interrogation, and X-ray and millimetre-wave technologies are supposed to displace the inquisitive physical 'touch' of the security operator (Amoore and Hall 2009). The emphasis of the more caring touchy-feely welfarism of the DHS resonates with the wider affective geographies that scholars such as Massumi (2005a) have pinpointed within contemporary apparatus of securitization (see Chapter Seven). Yet, the 'checkpoint evolution' materially spatializes these ethics in a new architecture of intensity and suspicion. For rather than simply replacing the 'risk-centered layered approach' of the TSA with 'human-centred design', they are now more strategically merged. The TSA's invention is a self-confessed effort not simply to calm the checkpoint for calming's sake, but to do so in a manner that incites and reveals anomalies and rogue patterns of behaviour, following in the footsteps of project 'Hostile Intent'.

Obsessed with distinguishing normal from 'hostile' intentions and behaviours, the TSA believe a calm checkpoint will make 'potential threats stand out'. Both 'security and the passenger experience are improved by making hostile intent more visible'. In this remarkable twist of logic, the very technologies which have been so vehemently criticized for reducing our quality of life, experiences of aeromobile travel, dehumanizing the terminal (Mutimer 2007), and relieving passengers of their freedoms, are now aligned with improving the atmosphere and feeling of security so that travel may even be 'fun'. As the TSA (2008a) blog explains:

> We'd like to change the environment for two reasons. First off, passengers deserve to travel through our checkpoints without feeling stressed. We all have enough stress in our day to day lives don't we?
>
> Secondly, the more relaxed passengers are, the easier it will be for our Behavior Detection Officers to pick out folks who are displaying signs of fear and deception.
>
> In a relaxed environment, someone with ill intent will stick out like a man in a plaid suit at the Oscars.

The tones, hues and colour spectrums of the DHS's affective modulations are now flattened down materially and horizontally in the cooling blue

Figure 5.4 The new smiling TSA screener. (http://www.tsa.gov/evolution/innovation.shtm. Reproduced with permission of the Transport Security Administration.)

colour palettes of Baltimore-Washington International Airport. Passengers will now face 'light panels emitting "calming" colours while soothing, ambient "spa-like" music will be played on loudspeakers' (TSA 2008a). Ironically, passengers now come face-to-face with the faces of their security counterparts. The TSA hope that biographical boards will disclose the lives of the screener staff, their expertise, their efficacy and successes in catching 'bad guys'. All this 'will put a face on and show the personal side of our screeners' (TSA 2008a). The TSA's continuing strategy of 'passenger engagement' even sees the smiling screener embodied in avatar form on their upgraded website and able to be spun 360 degrees to identify their new and vividly blue uniform (Massumi 2002; see Figure 5.4).

The TSA's 'evolution' is self-proclaimedly addressed to the *environment* of the passenger, overlaying their field of suspicion onto a horizontal space of norms and appropriate behaviours. The environment is an intentionally soothing one. De-composure and re-composure benches are new ergonomically designed features giving passengers the chance to reassemble their belongings after scans. They are meant to be social, facilitating interaction and friendly contact between passengers. This environment is also aimed at a region of the passenger which is much more primordial than cognitive. The checkpoint employs clues, triggers and action-points which require little thought or passenger deliberation, removing the possibilities of confusion. TSA public videos emphasize how these triggers are not one-way directions or orders, but are rather 'invitations', 'prompts' to action (TSA 2007).

These efforts have not gone unnoticed or without critique. One entrant to the TSA's online blog takes considerable offence at their efforts subtly to influence behaviour, seemingly aware of the Spinozist cartographic mapping of passenger and terminal space to a schematic of impulses and attractors. With much annoyance, the blogger describes how the triggers assume some sort of reduction of the inhuman body:

> I do not need a sign to 'prompt my behavior', I am not a lab rat. Give me signs that guide or inform. I know you are thinking 'means the same thing', you are wrong. You prompt behavior on people that do not know how to behave like criminals. We are NOT criminals, we are your bosses and your fellow citizens, treat us as such. Think of us a such. (TSA 2008b)

What we are seeing in this example is a peculiar attunement of American airport security services to the *milieu* they are dealing with, constructing new and environmental fields of normality and calmness.

Attention

We have examined techniques which attempt to improve the propensity to spot and distinguish instinctive and reactive behaviours – quite different kinds of gaze to those discussed in the previous chapter. These forms of seeing by the 'gut' can be developed through habituated and automated forms of training which improve attention.

A pilot's or observer's ability to see or act under the trying conditions of night flying, or the avoidance of bullets and flak and bad weather, has been a common problem. During the Second World War, this issue was brought to the surface in the difficulties experienced by pilots and their observers in identifying enemy aircraft when 'high altitude, the physical and informational environment of the cockpit, and the respiratory and central nervous systems of the pilot all combine to modulate the critical attribute of attention' (Hookway 2006: 37).

Bartlett's Cambridge studies found that the sorts of 'proprioceptive impulses which would normally be set up by movement of the machine' and given to an alert operator could be interpreted as useful information. For the tired pilot, however, they began to be seen as 'unwanted and disagreeable stimuli of bodily origin' (Bartlett 1943: 253). Adaptation to flying at night was an issue that plagued pilots as they shifted from illuminated lights or luminous paint within the cockpit to the blackness outside. Pilots adapting quickly from the bright conditions of the cockpit to the outside darkness would prove vital in both safety and aerial terms. Finding that red light scarcely affected adaptation, red illumination was tested using a 'mock-up' instrument panel in the Cambridge Cockpit. When the cockpit needed to be illuminated, red light proved even more useful. Measured by timing the pilot's visual threshold of adaptation once the illumination was switched off, red illumination was found to be 2.5 minutes faster than normal light.[53] The aircraft cockpit quickly took on the red hue of adjusted illuminated instruments.

Skilled adaptation coincided with the critical attribute of the aerial body to come to and sustain its attention. Tests were devised in which observers heavily schooled in aircraft types were compared with relative rookies in their ability to identify aircraft silhouettes. The results were initially puzzling. Highly trained observers were actually handicapped in searching out enemy aircraft, for in 'picking out a large number of structural points' they had no time 'to go over all the important points and arrive at a reasoned conclusion', and 'may thus overlook essential points of difference'.[54] In their first experiments it was found that pilots were naturally divided into two camps of capacity: 'those who waited until they were certain of the nature of the subject at which they were looking, and those who made several guesses before they reached the right answer'. The matter, scientist P. C. Livingstone wrote, 'was not merely a visual one, it was psycho-visual'.[55]

Whilst the trained observer looked out for minute nuances of detail, the equally successful untrained observers selected a few important points to concentrate their attention on; these included variables such as 'number of engines', width of wing-span, shape of wing – tapering, parallel-edged, elliptical, pointed, rounded or blunt – tail shape, size of fuselage. Perception was made not with the sensitivity to all 'total perceptual patterns (Gestalten)' but by stressing the necessity of 'observing isolated details'.[56]

Similarly, the ability to form the accurate and quick judgements of spatial relationships and shapes was compared with the process of making an approach to land. Whilst an efficient pilot would 'unconsciously choose one or two points on which to form a judgment and will inhibit all other impressions', other pupils, attempting to take in all the information, would be 'quite unable to discriminate between them and reset appropriately'. The issue was that the pilots or observers needed to be able to inhibit impressions and other detail less essential to the task before them. Observation and discrimination could then be a process not of complicated interpretation, but of swift and immediate action.[57] In the civilian sphere, 'knowing *where* to look' for certain distinctive points and qualities was instructed by R. A. Saville-Sneath's (1941) popular guide. Following the pilot's success, the guide advised a recognizably familiar ability to distinguish aircraft with *certainty*. The 'instantaneous, apparently instinctive' recognition of aircraft is the 'finished performance, the final stage of proficiency to which all training is directed' (Saville-Sneath 1941: 7).[58]

In just this manner, establishing the identity of an aircraft overlaps with the spotting of faces discussed and our instinctive police officers with whom we began. To the trained spotter, the distinctive points of the aircraft are 'as easily recognizable as the features of a familiar face' (Saville-Sneath 1941: 7). The judgements could be helped along by ways of seeing or 'methods of looking', of which, Craik argued, some 'are better than others'.

By understanding the inner workings of the retina, Craik suggested that focused interpretation would be far less effective than quick scanning. Drawing upon evidence to suggest that a stationary image, particularly in night conditions, would rapidly become 'invisible', and that the sensitivity of the retina was also incredibly unstable, he concluded that '[i]t is pointless to scan in close tracks'. Focusing the attention of the eye to 'laboriously scanning some distant part of the field' would blind one's vision to the immediate foreground. A rapid return to the original direction of fixation could reveal a previously unseen aircraft. The eye's ability to pick out detail whilst in quick movement was more powerful than careful scrutiny. 'The period during which the eye is moving from one point to another should not be regarded as wasted,' argued Craik. For if the movement did not exceed around 5 degrees per second it could be remarkably effective in picking up silhouettes.[59] 'Slight eye movements at about 5 degree per second over a 5 degree path backwards and forwards seem to be most effective; they also permit the observer to distinguish between subjective bright and dark patches and external objects,' Craik concluded.[60] The skill to be developed was thus one of making 'making rapid hypotheses or guesses as to the nature of objects, and testing out these hypotheses', based upon ways of looking that were relatively swift and, importantly, unfocused.[61]

As we saw earlier, the detection of facial expression in order to identify intent relies upon a similar process of scanning and picking up swift and almost imperceptible movements. Contemporary security processes recover the same tension of the pilot observers. Today both behavioural and object-centred surveillance is being developed in airport security in order to monitor the flows and trajectories of dangerous objects, while also identifying threatening objects from innocent ones. Paralleling the behavioural profile, this is being developed through the increasing capacity of an agent – like the pilot – to 'thin-slice' (Gladwell 2005). X-ray operators and security personnel employ a form of visuality and inspection that extends beyond mere sight to foresight. When a 'cell phone can be a gun, a lipstick can be a knife, a teddy bear can carry a weapon, a condom can be a vessel for drug-running, and a shoe can be a ticking time bomb' (Parks 2007: 195), X-ray operators must discriminate one object from another by determining the possible from the actual. What an object may be once it has been stripped of its disguise, or, rather, what it may become. Reminiscent of Jordan Crandall's 'anticipatory form of seeing', screeners create an '[i]nclination-position [which] scripts an object that is already ahead of itself, a shadow future state that exerts a strong gravitational pull' (Crandall 2006).

Screening, sensing or scanning is meant to occur quickly. 'TSA workers rely on so-called "screener assistant technologies", such as "Target", that use software algorithms to search X-ray images for dangerous materials – the algorithm analyzes the mass, size, and atomic number of items in the image against preset thresholds; objects that match the defined criteria are identified for the operator' (Parks 2007: 190). In order to train security personnel to perform these complex tasks of inclination/foresight quickly, routinely and effectively, screeners are required to undergo continual periods of training on simulated X-ray machines. Software simulations are used in order to augment practices that monitor and regulate the mobilities of humans and objects through the airport.

Certain objects and behaviour (e.g. unattended bags or apparently 'shifty' behaviour) are traced and identified as threats to the airport and, by association, the nation. X-ray Tutor, a software simulation employed by screeners of the Canadian Air Transport Security Authority (CATSA), enables prospective as well as trained screeners to learn how to identity different threat-objects within the images that are projected onto their display screen. Screening officers are able to 'play the game', developing their proficiency at being able to identify forbidden items as well as potentially harmful combinations of otherwise 'safe' objects. To train and improve rates of detection and differentiation, screening officers can tailor learning programmes to individual students.

X-ray Tutor can purportedly decrease the time taken for a screener to identify a threat from 8 to 4 seconds, a significant reduction when hundreds of bags require processing. Drawing on research in psychological, cognitive and human–computer interaction, computer-based training (CBT) systems (of which X-ray Tutor is one) work by improving visual memory recall, replacing on-the-spot cognition and calculation with a substratum of intuitive recognition (Schwaninger and Hofer 2004). Training individuals to recognize threats, as well as around 70–80 permutations of an object varied by rotation, dimensions and colour, the system enables instinctive judgements to be made that require little thought and deliberation (the so-called 'practised eye' of the screener), relying more on 'visual memory representations' than on 'increased general visual processing capacities' (Schwaniger and Hofer 2004: 156). Like the stereoscopic impression, this was a way of viewing that was fast – an 'instantaneous illumination' such as an electric spark.

Further developments have seen simulated threats *projected* onto a screener's display in real time. Of course, the basic augmentation of reality delivered by normal X-ray machines already simulates and stimulates the presence of a threat by colour-coding a bag's contents according to the sorts of materials detected within it. Thus, colour palettes and hues already induce a reactionary association between colour and threat that is rendered into disposition (see Massumi 2005a on threat levels). Furthermore, Threat Image Projection Software (TIPS) works to do more than shape the pre-reflective thought required to identify a threat (Hofer and Schwaniger 2005). TIPS demands a kind of bodily and affective posturing of the screening officer so that they are continuously alert to scrutiny and modify their own behaviour accordingly.

Just as seeing is occasionally associated with a rather distanced and distracted form of engagement (Friedberg 1993; Morse 1990), distractions reducing screener attention to a threat are being transformed according to shifting forms of security-software design. The continual aggravation of screener attentiveness is heightened by the knowledge that an individual's performance over the course of a shift is recorded – the expectation clearly being that they will continually want to better their 'score' and outperform their colleagues. Resembling a kind of Foucauldian self-discipline, screeners must assume that they are being monitored constantly, reinforcing what Parks calls a 'high alert banality' (2007: 193). Furthermore, they must be more than attentive. TIPS requires a state of 'alertness', where one is ready for an unknown but omnipresent and imminent threat.

The 'technological processes of foresight and anticipation' (Virilio 1989: 86) are thus essentially composed by 'subconscious visual reflexes' (Virilio 1997: 93).

Conclusion

Atmospheres of pressure; ambient forms of suggestion and control and the disciplining of attention; gravimetric fields and forces pressing on the body, provoking response – these were the concerns for the aerial body's performances and capabilities. Assemblages of objects and life exterior to the body were addressed. Temporary and semi-permanent environments were created which could test the body, improve it, allow it to adapt and encourage it to do something. These techniques were intertwined in helping the body survive violence, whilst they also inflicted it: the German terror on their *Versuchpersonen* being an obvious example.

In the following chapter and section, we dwell on the environment and other ambient forms of materiality as a similar sort of deliberate terror, waged not in order to improve the body, but to incur violence, to alter perception and will – directed towards the 'very human-ambient "things" without which people cannot remain people' (Sloterdijk 2009: 25). Whilst the *Versuchpersonen* were deemed less than life and duly acted upon under this presumption, aerial violence works to produce similarly non-living subjects through the attunement of aggression towards the environments that supply the conditions for the termination, or degradation, of a population's survival.

Part Three

Aerial Aggression

Chapter Six

Aerial Environments

During the years of World War II, American and British planners and analysts learned to see through a bombsight. Not in a single glimpse, but in the routine killings and losses that accompanied ever more frequently repeated raids. [...] A calculus of fractions, probabilities, delays.

(Galison 2001: 28–9)

[G]iven an animal, what is this animal unaffected by in the infinite world? What does it react to positively or negatively? What are its nutrients and its poisons? What does it 'take' in its world? Every point has its counterpoints: the plant and the rain, the spider and the fly. So an animal, a thing, is never separable from its relations with the world.

(Deleuze 1988: 627–8)

[M]ilieu does not evoke any relation except that of a position endlessly negated by exteriority. [...] The milieu is truly a pure system of relations without supports.

(Canguilhem 2008: 103)

Introduction

During the Blitz the British Prime Minister, Winston Churchill, visited Plymouth on 2 May 1941 after days of aerial bombardment had destroyed parts of the town. His tour was met with a crowd of around 120 people by the guildhall. Tom Harrisson, the director of Mass Observation, provides us with details of an observer's report. According to the testimony, Churchill's physical presence had an enormous impact upon the town's suffering: '[T]he experience was profoundly moving. He brought (we thought) a breath of positivity, vitality and above all *concern* into what appeared by then to be an "atmosphere of negative devastation"' (Harrisson 1979: 211). Whilst several public figures had addressed the people quite unsuccessfully, there was something about Churchill's comportment that made his presence as he moved through the crowd incredibly *moving* (Gregory 2009). It was his face. An observer wrote how he was 'fierce' and 'firmly balanced on the back of the car'. He had 'great tears of angry sorrow' in his eyes. The Plymouth mayor's wife, Nancy Astor, commented, 'It's all well to cry Winston, but you've got to do something' (cited in Wrigley 2002: 27–8).

Churchill's face did do something. His visible emotions displayed empathy, showing that he was 'moved' by the suffering he witnessed. Yet there was something else affecting in Churchill's expression. He was 'visibly determined to see that it spelt not defeat but victory'. His tears of sorrow ran alongside his 'fierceness' and rigid strength. Overflowing with 'human sympathy', he was described as an 'epic figure riding through this epic of destruction', determined and uncompromising (Harrisson 1979: 211).

Today the now rather clichéd representations of 'Shock and Awe' in the US bombing of Baghdad have marked what various scholars might call a kind of post-modern war (Hables-Gray 1997): a war mediatized by the communications vectors that transmit its power outwith to be consumed, and symbolized by the perpendicular intersecting lines of the digital bomb-sight (Kaplan and Kelly 2006). It is the zenith of networked warfare made possible by a combination of informational and simulation techniques that allow air strikes to be planned, rehearsed and coordinated with split-second accuracy and at lightning-fast speeds (Der Derian 2001). Moreover, it is a demonstration of what are now labelled 'effects-based' operations. Where military airpower is deployed in order to achieve political gains, 'rapid dominance' deploys the 'stunning' use of aerial force (Ullman and Wade 1998).

In both contexts we have affects and effects.

Like many leaders, Churchill was a master communicator and conveyor of feelings – he was charismatic (Lindholm 1990) – and his intervention into the despondent population of Plymouth marked an interruption of a field of affect. His face did it; his path through the destruction was a path through the material and psychic remains of a city undone. Notably altered by what he saw, Churchill acted back upon the environment through which he moved, modulating the atmosphere, if only for a while. Not nearly as surgical as its media and military proponents claimed, the bombing of Baghdad was also Churchill's strategy. Falling with the power of Churchill's tears, the missiles of operation Iraqi Freedom rained upon the feelings and *will* of the Iraqi people. They were transmitted through atmospheres and material devastation, and they rehearsed the RAF's strategic terror first waged during colonial occupation. From the early beginnings of colonial policing in Arabia to the doctrine of Arthur Harris' strategic bombing in the Second World War (Hastings 1979), to today's 'effects-based' operations

(Carey and Read 2006: 63), airpower has been about determining affects through the explicit targeting of environmental effects.

This chapter is concerned with the aeroplane's production of environments, not necessarily *the* environment but the *environments* it imagines, effects and reaches through. Focusing on the utility of the aeroplane as a medium of war and terror, the chapter will explore how the *environment* has emerged as a key medium or terrain through which aerial violence and securitization can be waged. As opposed to the direct touch or interaction with the subject or the population, I argue that we are seeing an apprehension and address of what Foucault might call the biopolitical management of the milieu.

The milieu is 'what is needed to account for action at a distance of one body on another', a medium 'in which circulation is carried out', upon which action takes place, 'a certain number of combined, overall effects bearing on all who live in it' (Foucault 2007: 20–1). What could better describe the mechanism of the long-range bomber as an attack upon the medium of life, a direct hit not on the subject but upon the conditions necessary for a population to survive (M. Davis 2002)? Following Peter Sloterdijk's (2009) assumption of aerial bombing as an address of the conditions and the capacities essential for one to live, I argue that it is the indirect effects upon the population's 'vital' relationship with its environment that mark out the affective and effective capabilities of aerial war and terror.

The Emergence of a Target

Know your enemy. Such was the wisdom of the first influential theorists of airpower. Envisaging a set of vulnerable targets endowed with capacities to affect and be affected, lessons learnt from the experiences of colonial policing were deposited into a sediment of doctrine and practice by influential figures such as Hugh Trenchard (Meilinger 1996; Paris 1989). During the formative years of the RAF as an imperialistic military power, the aeroplane was directed towards 'air control': the task of policing the Arab populations in Iraq, Palestine and Afghanistan (Omissi 1990; Satia 2008).

Unthinking aerial war

Beyond the powers of orthodox military thinking, the British had learnt to *unthink* in order to understand the Arabian mindset. Sidelining the usual empirically driven intelligence gathering of a region and its people, the uniqueness of the landscape and the population posed new problems for military strategy. If the Arab jettisoned intellectual and scientific thinking in favour of wisdom and intuition, so did the British. Seeing the purported influence of the landscape on the Arabian people, Priya Satia (2006, 2008) notes how the 'Arab wisdom' 'was as much a product of place as of race'. In order to grasp 'Arabia's political and geographical realities', the intelligence strategy sought to conduct 'lengthy immersion as the only effective preparation for this work'. Quite unlike the immersive experiments conducted on the Dachau prisoners, the agents would '*think like an Arab*, an empathetic mimicry of the Arab mind', by copying the patterns of their thought. The agent Captain Norman Bray sought to 'merge [...] in the Oriental as far as possible, [to] absorb his ideas, *see with his eyes, and hear with his ears*, to the fullest extent possible to one bred in British traditions' (cited in Satia 2006: 205). Through this kind of 'sympathetic projection', the population could be known, understood and therefore controlled.

The British intelligence forces draw out familiar conceptions of the nomad subsisting with the land. But unlike the state in *A Thousand Plateaus* (Deleuze and Guattari 1988), the efforts of the British are directed towards their own immersion in the desert theatre. 'There appears to be a sort of natural fellow-feeling between these nomad Arabs and the Air Force,' remarked Robert Brooke-Popham the RAF's director of research, perceiving a basic 'congruence between the liberty of action of the aircraft and the desert warrior', both operating in empty, unmapped, magical spaces (Satia 2006: 13). T. E. Lawrence, who had researched 'Bedouin warfare' for an alternative to the anonymous mass slaughter of the Western Front, prophesied, 'What the Arabs did yesterday the Air Forces may do to-morrow. And in the same way – yet more swiftly' (cited in Satia 2006: 13). Although the targeted subject changed in the theatre of the Second World War, an understanding of the population in relation to their environment was sustained by the political and military experience of trench warfare during the First World War and the Spanish Civil War (Patterson 2007; Sloterdijk 2009).

Military intellectuals and leaders of the inter-war period anticipated what a full-scale air war would look like, and how it would impact upon everyday life (Lindquist 2000; Townshend 1986). Born out of the experiences of the First World War, the idea of the civilian and the 'city-as-target' took hold (Biddle 2004; Bishop and Clancey 2004; Grayling 2006). Popularized and naturalized further, understandings of the aeroplane's violence were perpetuated by Hugh Trenchard, Italian strategist Gulio Douhet and British military intellectuals such as L. E. O. Charlton, Noel Pemberton Billing, J. B. S. Haldane and Basil Liddell-Hart. From these views, *morale* surfaced as a strategic object, summarized by J. M. Spaight as a 'disturbance of life, the dislocation of routine' – 'the moral in effect is far more important than the material' (cited in Biddle 2004: 71).

Douhet's *Il dominio dell'aria*, first published in 1921, argued that the aeroplane would bring war to the civilian. From the zeppelin attacks of the First World War, predictions rained of utter terror and frenzied mobs (*Lancet*

1917a, 1917b; Waller 1918). Public figures foresaw the apocalyptic situation of 'ruined cities [...] hospitals filled with the maimed and mutilated of all ages and of both sexes, asylums crowded with unfortunate human beings whom terror has made insane' (Spaight 1938: 48). An image of uncontrollable panicky mobility or immobile paralysis emerged. Oliver Stewart, writing in the *Morning Post*, declared that the 'doctrine of central shock is simply the doctrine of the bomb' (cited in Spaight 1938: 43). The Douhetisian view was expressed by Winston Churchill in the House of Commons in 1934, imagining an uncontrollable 'vast mass' of people who, without shelter, food and sanitation, would 'confront the Government of the day with an administrative problem of the first magnitude'.[1] The air war generated its own type of fear, yet what was really feared was not the actual attacks, but the effect of the fear of the attacks. James Kendall's book *Breathe Freely!* described the threat of poison gases dropped from the air: 'The danger of fear, of course, is that it may lead ultimately to panic, and panic is really the only thing that we have to fear' (1938: 79). The air war was almost overdetermined. Apprehension came from the projection of its complexity and ignorance over what it would mean.

Promulgating the Douhetsian view was a combination of influential military thinkers drawing on experience of the first uses of airpower, psychologists attempting to make sense of the events of the First World War, and military practitioners who attempted to do both. The targeted urban became a maelstrom of feelings (Thrift 2005a), 'a vortex of broken passions, lusts, hopes, fears and horrors' for D. H. Lawrence (cited in Patterson 2007: 86). Indeed, it was these passions that came really to shape and justify bombing strategy as well as its defence (discussed in the following chapter). Vengeance, counter-strikes and not much 'else than instinctive opinion' that 'wide spread neuroses and panic would ensue' became the status quo (Titmuss in Biddle 2004: 80).

Gustave Le Bon (1925) saw the Great War as a psychological opportunity, a 'vast laboratory for experimental psychology'.[2] In the convergence of opinions and diverse expertise, 'morale' was constructed as an object that would become almost central to the next 80 years of aerial warfare. In the British context, Brooke-Popham, who led the RAF's research division and later became Commandant of the RAF's Staff College, identified morale as of 'paramount importance'. It was 'essential for all officers, especially those on the staff, to make a study of it'. To do so was not easy given the debilitating lack of research in the matter. There exists, Brooke-Popham explained, 'a natural reluctance on the part of all Englishmen to talk freely about human instincts and emotions'.[3]

Drawing on biological and ethological theories on reproduction and human nature, Brooke-Popham identified certain physiological characteristics or bodily infrastructures for the production, reception and maintenance of morale. These were the instincts, which, under times of stress, would govern behaviour rather than reason. The greater the stress experienced, 'the more a man does tend to revert to his primary, and therefore most deep-seated, instincts'. Of this substratum supporting the expression of morale and action, Brooke-Popham identified three main characteristics. The first of these were 'basic', or primordial – self-preservation being one. Secondly, with the 'instincts impressed by method of life' across animals and humans, Brooke-Popham described the group instincts – the preservation of something larger than one's self. And thirdly, there were those instincts differentiated across different races or classes. These, Brooke-Popham found, were, 'a) Fanaticism for religious dogma; b) Love of dogma; c) Desire to overcome difficulties and to face dangers' – a quality he saw evident within the 'Nordic races'.[4] Fighting man, he argued, had to be understood in terms of his 'human nature', an essential knowledge for 'anyone who aspires to success in war'.[5]

In all three types of instincts, a population expressing good or bad morale (collective affect) was viewed as generally regressive and susceptible to its environment – be that material, biological or social. Racial instincts appeared to Brooke-Popham 'as important as a part of intelligence duties as that of the strength of navies and armies', and developed through Lamarckian evolution by interplay with the environment. The preservation of the group, with assistance from Le Bon's psychology of the crowd, was a demonstration of the individual's sublimation into a socially supportive milieu, where 'each one tends to lose his individuality, and the whole mass becomes, as it were, self hypnotised'. This is a mass who 'is swayed by impulse and emotion, not by reason, and tends to look about for, and to place itself under, a leader'.[6] Even the leader is not an individual, just perhaps the most susceptible to the influence of the crowd; they 'mutually react on one another'.

Brooke-Popham's influence at the Staff College cemented 'morale' as the key to unlock the application of leadership, as well as the effectiveness of airpower. *Esprit de corps*, enthusiasm and the necessary discipline to sustain these collective affects (in other contexts see Forster 2003; Harrisson 2000; McNeill 1995; Moran 1945) would ensure the most efficient behaviour for an aerial force. Enthusiasm, a facet of morale, was understood to arise from emotion; it was transient and unstable, 'and so requires to be continuously stimulated'. It developed a 'desire for action', 'seeks some definite focus around which it may crystallize', and the 'higher the ideal which forms the cause of enthusiasm, the stronger and more durable will that enthusiasm be'.[7]

Understood in this way, the officer could be immersed in morale and have his thoughts and plans of real operations structured. For instance, Brooke-Popham set an exercise pertaining to the logical sequence of a bombing operation. The officer had to rearrange the sequence in order to plan a successful raid operationally in terms of the targeting of morale:

a) The factory is unlikely to be located at night unless there is sufficient light, i.e. unless the attack takes place two hours or more after the moon has risen.

b) The factory must be bombed while the men are at work, otherwise the effect will be negligible.

c) On the 19th the moon rises at 2100, it will therefore have been up two hours by 2300, but every night it rises an hour later.

d) The only aeroplane we possess that have sufficient petrol capacity to reach the factories are our heavy bombers. These can only operate at night i.e. between 1800 and 0600.

e) There are men at work in the factory between 2300 and 0700.[8]

Other members of the cadet staff, college courses, exercises, essays and conferences cultivated the belief in the benefits of understanding 'morale'.[9] Altogether morale was solidified as a strategic object, for maintaining a buoyant, determined and disciplined air force, and as a target against which those forces could be directed.

Classquakes

Thinkers who witnessed the 'stampedes of threatened populations', as named by Frank Morison's study of air raids on London during the First World War (Morison 1937), wrote of the expected 'undisciplined flight', resulting in a self-evacuated population in a state of semi-starvation and lawlessness. In these predictions we find a class politics of affective events that are well described by Kenneth Hewitt's (1983) term *classquakes*. For L. E. O. Charlton, the factory and working classes would demonstrate the lowest forms of behaviour; they would prove more 'difficult to control' and 'more susceptible than most to dismay and stampede when the air-raid warning goes' (1935: 173). Similarly, Quentin Reynolds writes that in the first few weeks of the Blitz, 'men and women lived like hunted animals. They ran towards shelters when the bombs began. They crept out when the immediate danger had past' (1941: 65). The more animal-like the population became, the easier they were to predict:

> It is known that there is a section of the population, estimated at a maximum of one-tenth, who are of a weaker constitutional mental makeup than the rest. These people react to different situations in two ways – either by a cowardly retreat, or by a neurotic mental breakdown. [...] The potentially neurotic section of the population takes to the roads each evening and seeks safety in numbers. (McKay 2004: 81)

This group were considered to have a 'bodily inertia induced by unemployment'; they would be quick to panic and react to an air attack and the first to be rendered shocked and immobile – in other words, 'a decided danger at the outbreak of war'. Moreover, the togetherness of such a population posed further problems for the transmission of feeling. Charlton predicted that a 'widespread alarm and despondency' would shoot through the country at 'bullet speed', while 'rumours of disaster everywhere' would be 'repeated to excess' (1935: 129). On the other hand, the middle classes were more 'inured to civic discipline and obedience'. In their fractured individualism they lacked the herding and group dynamics of a mob – what Charlton describes as the 'cohesiveness necessary to law-breaking in the revolutionary sense' (1935: 129).

Whether regression into an infantilized state of incapacity or possession by an overwhelming urge to flee, the anticipated effects of aerial bombing drew upon popular theorizations of affect, emotion and mobility by writers such as Gabriel Tarde and Wilfred Trotter.[10] For Trotter, war and its accompanying stress and apprehension could invigorate a population with feelings of loneliness and isolation that could only be quelled by an active desire for company and physical contact. Yet bringing populations together allowed social feelings to spread rapidly between members of the herd; the 'thrill of alarm' could propagate with 'magic rapidity' (Trotter 1924: 143). In this thesis the herd instinct caused issues for panic, rioting and insurrection. Feelings of fear and hostility, according to Trotter, harden the character of a herd and increase its feeling of vigour and resistance. Trotter claimed 'that in a country at war *every* citizen is exposed to the extremely powerful simulation of a herd instinct characteristic of that state' (1924: 216; on defeat see Schivelbusch 2001).

Catatonic shock featured as another excess of aerial war. What became known as 'shell shock' or war neurosis (Babington 1997; Holden 1998; Leese 2002) was of particular interest to the British government, who commissioned a major report in order to understand the problems of the battlefield.[11] Against the background of an industrial, disciplined and productivist capitalist economy, both shell-shocked soldier and working-class civilian were viewed with suspicion as malingerers (Bourke 1996; Shephard 2001). Under these terms the traumatized soldier was a body similar to the striking factory worker, becoming 'frozen' and unproductive (Griffin 1993; Jones and Wessely 2005). As interpretations of shock remained quite diverse, discord emerged around what could possibly cause it. For some, shell shock could be put down to the physical encounter with an exploding shell. The resulting concussion, it was thought, would cause many tiny haemorrhages in the brain, with the result of severe damage to the mind and loss of bodily control. Others believed that an overexposure of the body to all the environmental triggers of war would wash over and through the body, attacking the soldier 'from within and without' (Myers 1940).

Calculation, amplification and the geographies of an explosion

In the context of the above, the work of two key advisers and research scientists to the British government sought empirical solace in the volatile materiality of the explosion. Their mapping of blast's intensities, peaks and amplitudes of inanimate and biological matter sketched out the micro-geometries of much larger detonations and maelstroms of urban affect. Through these logics, Derek Gregory argues that war was relocated to 'an abstract rather than an affective space' (2007: 18).

From the biological sciences, Solly Zuckerman and J. D. Bernal conducted influential work for the Ministry of Home Security, providing critical advice for the strategy of air-raid shelter design, the maintenance of morale, and the Allied offensive campaigns (especially the moral bombing attacks on Germany). Many other promising scientists from the natural sciences were also enrolled to apply their expertise at that time. Conrad Waddington (1973), for instance, applied evolutionary predatory behaviour to the detection and hunting of German U-boats by Allied aircraft.

Zuckerman had come from a distinguished career as a primatologist and endrocrinologist. His work was concerned with the relation between primate nervous systems and their outward behaviours of skill, imitation and reproduction. Of imitation, Zuckerman's study convinced him of the monkey's susceptibility to the movements and stimulations of neighbouring animals. Rejecting the anthropomorphism that underpinned ideas of the monkey's conscious mimicry, Zuckerman recognized the social and material setting of the ape's social life as an *environment* (Zuckerman 1932), continuously modified by the activities of others (resembling Trotter). Imitation was thus understood as a form of 'behavioural adaptation'. The behaviour of one monkey modifies the behaviour of another by 'constantly introducing into its immediate environment elements and relations that were previously unnoticed' (Zuckerman 1932: 171). It was this relation between the surface, the internal and the outer environment that came to characterize Zuckerman's research on aerial bombing.

Whilst reticent to apply human characteristics to the physiological behaviour of the primate's nervous systems, from this intellectual background Zuckerman set about conducting research for the Home Office examining the effects of blast concussion waves and the operation and possible defence of dropped shells. Medical practitioners would find 'little essential difference' between the findings Zuckerman found in his animal test subjects and bombed human casualties (Hadfield 1941).

This relation between the inner and outer worlds of the body was of 'special significance' for Zuckerman. Bomb causalities often displayed 'no visible outward signs of wounding", and yet 'the tiniest fragment of a projectile had produced extensive damage inside the body' (Burt 2006: 305). This mysterious phenomenon created a new terrain of problems and a new set of *conditions* or 'circumstances' through which the population had to survive:

> More tons of bombs are dropped on his country almost every night than were dropped in the whole of the 1914–1918 war. They are dropped not only in any one place, but far and wide. To-day everybody is exposed not only to the health risks of peacetime, in many cases intensified by changed living conditions, but also the casualty hazards of air raids. (Zuckerman 1941: 171)

To deal with these new circumstances and potential casualties, Zuckerman advocated the birth of a scientific sphere of research 'which demands in its attack more than simple clinical observation'; it requires empirical observation and simulation of 'the actual circumstances under which different types of casualties occur' (1941: 171). To these problems Zuckerman introduced an array of controlled experiments on animals which he subjected to concussion-causing blows, blast waves and fragments of shrapnel.

The shell blast had proved itself a disturbing urban materiality which troubled the Home Office as they attempted to mitigate the effects of bombing and increase the resilience of the bombarded population (Crichton-Miller 1941).[12] Knowing accurately the material and moral effects of blast could create an effective weapon and an effective means of defence. In one of his many reports Zuckerman explained the notion of the blast wave. By blast, he wrote,

> is meant the compression and suction wave which is set-up by the detonation of high explosive (H.E.). At every point in the neighbourhood of an explosion there occurs first a momentary wave of high-pressure (for about 0.006 se. for a 70lb charge), and then a negative 'suction' pressure, due to the fact that the positive compression wave has reduced the density of the air behind it to below normal atmospheric pressure. (Zuckerman 1940: 219)

Blasts were imagined and experienced both inside and outside the scientist's lab. With the population's peculiar and heightened sensitivity to the objects and materials that made up the urban environment, Johanna Bourke takes the example of a young girl and her brother running back towards their house: 'The oddest feeling in the air all around, as if the whole air was falling apart quite silently', 'waves buffeting me, one after another, like bathing in a rough sea' (cited in Bourke 2005: 223).

A medium-sized bomb would create a blast that travelled around 1,200 feet per second. Appearing in many testimonies of Blitz experiences, people described being pulled back, carried along by a tidal wave of air. It was

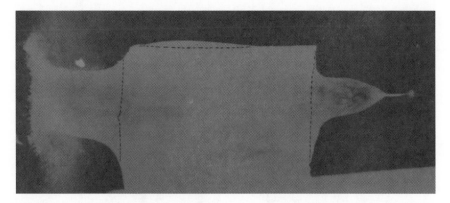

Figure 6.1   A rectangular gelatine block 207 microseconds after impact. (A. N. Black, B. Delisle Burns and S. Zuckerman, 'An experimental study of the wounding mechanism of high-velocity missiles', *British Medical Journal* 2 (1941): 872–4. Reproduced with permission of the BMJ Publishing Group Ltd.)

apparent that the split-second experiences of wind and energy required sufficient description and analysis. A body as large as a human being, wrote Zuckerman, 'would certainly be completely immersed for an instant in a wave of almost uniformly raised pressure' (1940: 219). All things in the immediate neighbourhood of an explosion 'will thus first experience a violent increased wind and hydrostatic pressure, which may tear them to pieces and blow them far from the scene of the explosion' (1940: 219).

A new molten urban, unbuilt and in motion. The raid's injuries, moreover, were strange and unknown. For people exposed to an explosion, flying debris of bomb fragments, bits of road, buildings – any material exploded by the blast – was particularly lethal. But exactly how blast injured the body was not fully understood as injuries appeared amplified – out of all proportion to the forces unleashed. Why, for instance, did tiny fragments of flying shrapnel cause such enormous damage within the body, such as broken bones and huge internal damage to flesh and vital organs? It appeared as if the injuries were being made from the inside out. Designing experiments to see the movement of shrapnel through gelatine blocks and the hind legs of rabbits, Zuckerman's team used shadowgraphs to photograph the impact and effect of the shrapnel damage, capturing the various stages of entrance, internal transit and exit (Figures 6.1 and 6.2).[13]

What they found was startling. After initial perforation of a gelatine block, the gelatine would undergo 'considerable expansion, until it becomes some three to four times its original volume',[14] after which the distortion of the block disappears and it returns to its original state of size and shape. The 'only permanent visible effect of the shot which is seen is the small threadlike track made by the passage of the ball',[15] as if a needle had been deliberately threaded through the material. After more experiments on the block

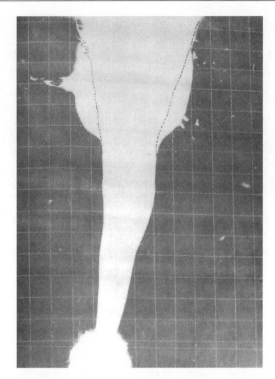

Figure 6.2   Rabbit's hind limb taken 550 milliseconds after impact. (A. N. Black, B. Delisle Burns and S. Zuckerman, 'An experimental study of the wounding mechanism of high-velocity missiles', *British Medical Journal* 2 (1941): 872–4. Reproduced with permission of the BMJ Publishing Group Ltd.)

and rabbit limbs, the team deduced that the expansion of the block, the limb's flesh and the bodies of people previously hit by shrapnel had undergone a process of *cavitation*. They argued that as a high-velocity missile passes through the block, 'it imparts motion to the particles in its track, and these fly off radially, imparting their momentum in turn to further particles (and so on)'. Occurring with considerable 'explosive violence', the process created a 'central cavity, the pressure of which is presumably subatmospheric, is formed within the target'.[16]

Analysing the distorted surrounding materials or tissue, Zuckerman's team likened the action to one of an 'internal explosion'. Minor atmospheres or volumes of air were formed within the body, deforming and stretching tissues around the central cavity after the missile left the region.[17]

Propagating the energy of the missile, the small yet disastrous 'internal explosions' of the projectile are effects of propagation and amplification. 'Under such conditions it is obvious that structures at great distances from the track of the projectile can suffer damage,' Zuckerman's team concluded.[18]

The problem was similar to that of blast, where even more diffuse and indirect kinds of damage were witnessed in the bombed body. Turning back to blast, the question that concerned Zuckerman was: why had so many people died when the victims showed no sign of bodily trauma, at least on the surface? One popular theory suggested that the intense waves of pressure which could follow an explosion would literally suck all of the air out of the body's respiratory system, reducing lung alveolar pressure sufficiently to cause lesions.

Zuckerman carried out various clinical experiments on several small mammals (these included cats, rats, pigeons, monkeys and rabbits) which were subjected to varying quantities of explosion at various distances from the explosive charge, and sheltered by layers of clothing and other materials. No animals showed any external signs of injury to their skins or outsides. By X-ray or autopsy quite a different picture emerged. Far from acting equally on the body from all sides, the experiments found that animals would sustain the most damage to one side of their body. If they were placed side-on to a blast, one lung might be damaged with the other unscathed, the solution almost certainly being that 'the side facing the explosion shielded the other from blast' (Zuckerman 1940: 222). The conclusion Zuckerman came to was that the trauma evidenced must be the action of the blast wave on the body wall. Further experiments showed that it was 'the pressure wave of blast which bruises the lungs by its impact' (Zuckerman 1940: 223). This pressure wave could be further prevented or diminished if a body was clothed in the right sort of material which could absorb and disperse the wave. From these findings, certain suggestions for medical diagnosis were possible.

Zuckerman's team concluded that the experimental blast injury was very unusual. It was a result of the *direct* action of blast upon the body, whereas in the outside world one was more likely to be injured or killed by other, more *indirect* means of an explosion. The environment or the *circumstances* in which blast acts 'are very unlike the experimental conditions that have been described' (Zuckerman 1941: 182). Listing these, Zuckerman describe the 'splinters and secondary missiles' from the bomb casing, the damage caused when 'houses collapse as a result of ground shock-waves', and the 'many other structural effects' that 'conduce to injuries'. Environmental hazards could also include the flame of the explosion, flying masonry, wood, glass and girders (Zuckerman 1941: 182). With projectile matter flying off radially like the shrapnel in the rabbit's limb, the casualties of an air raid were likely to be caused by violent indirect means.

Such studies provided useful empirical evidence for medical diagnosis and possible treatment. But they had more important consequences for understanding the process of aerial bombing, scientifically perpetuating the analogic and affective amplifications of morale and panic through the trope of the explosion and the body's susceptibility to indirect environmental effects.

The Ministry of Home Security investigated the patternation of indirect causes and consequences and how they might be mapped and mapped onto potential or future events.[19] Blast pressure was calculated to fall off accord-ing to distance. With a 70 lb charge, pressure would fall from 110 lb at 14 ft, 60 lb at 18 ft, 15 lb at 30 ft, down to 6 lb at 50 ft (Zuckerman 1941: 172). The exploding bomb would form a zone of danger geographically delimited to the borders of the *circumstances* in which it fell. Thus if a bomb fell close to a structure such as the walls of a building, 'the blast is reflected and its energy absorbed in odd ways, so that people situated near-by may escape the impact of the wave'. A geography would both shape and result from this violence. If a bomb was to explode after it had penetrated the ground and formed a crater, the resultant explosion would be directed by the shape of the crater architecture upwards into an expanding cone.[20]

J. D. Bernal made similar statistical surveys on the space of the air-raid shelter and the city of Coventry. The 'probable effects' of bombs dropped over specific targets could be estimated and protection directed towards maximizing the safety of the population.[21] Measuring the probable error of a bomb versus the amount of bombs dropped, the vulnerabilities of certain spatial structures could be calculated and, thus, the average number of cas-ualties a raid should produce would be found.[22] Bernal built up a picture of the bombs probable to affect spaces and cities, spatially projecting the vul-nerabilities of Coventry's factories with casualty rates.[23] Knowing 'the exact circumstances of their occurrence', it would be simple enough to predict the likely outcome of future events.[24]

Pulling up and out of the explosions and the inner worlds of tissue and bone, Zuckerman and Bernal addressed the impact of bombing on the wider organism of the city and its life, constituted by material productivity, mobility and collective sentiments of morale. Through analogies of arteries, circulations and systems, theirs was an address not to the 'human body [...] but the body politic' (Gregory 2009; see also Overy 1981), the intention of these wider studies being that if more detailed research into the 'subsequent effects' of enemy bombing could be known, then it would 'enable an opin-ion to be formed as to the possibility of bringing about a decisive break-down in morale in urban districts by air attack and that it will assist the Air Ministry to estimate the nature and intensity of the attack required to achieve this purpose'.[25] Understanding the country's own reaction to bomb-ing could help formulate the RAF's bombing strategy.

Forming alongside wider government efforts to take continual snapshots of the population's state of collective feeling, such as weekly Home Intelligence reports and Mass Observation, the aim of Zuckerman and Bernal's work was to measure and see morale (see also Swann and Aprahamian 1999). A slip-pery object to pin down or articulate, '[a]ny statement about the effects of air raids on morale' being 'necessarily vague', how to *see* morale was a problem. The photographic journal *Evidence in Camera* even remarked that 'there are many questions which PI [Photo Interpretation] could never answer, enemy morale being an obvious enough case'.[26] Yet for Zuckerman and Bernal, morale was inextricably linked to the explosive material process of the shell blast they had studied in the body. A specific component of the study on morale and the industrial effects of enemy bombing was to 'enquire into the possibilities of bringing about a decisive breakdown in morale in urban districts in general as an object distinct, though necessarily related to, material damage'.[27] Zuckerman would be asked, 'how many tons of bombs does it take to *break* a town?' and 'in what ratio should the total load be distributed and over how many nights?' (Zuckerman 1978: 141).

In their study of Hull, Birmingham and the Allied bombing of Tripoli, Zuckerman's and Bernal's reports articulate a scalar upshift of explosive material and affective energy. Zuckerman and Bernal were convinced of the depletion of Tripoli's morale and industrial effectiveness, by witnessing its inhabitants rushing outwardly beyond the city, the motion of the population simulating the people's emotion. Homelessness, trekking and irruptive volatile movements out of the city were easy indicators of successful bombing raids.[28] Mobility and behaviour expressed the efficacy of bombing on collective feeling.[29] The material and morale effects of an air attack could be out of all proportion to their instigator, just like the catastrophic internal injuries of shrapnel. A 'relatively low concentration of bombs' might 'completely demoralise an Italian population'. The momentum of the mobilities of evacuation and trekking represented a 'social upheaval far greater than any we have experienced as a result of air raids and far greater than has been reported as a result of our attacks on German targets'.[30]

The population's spatial expansion represented 'a violent social reaction to bombing'. Closely following the outline of the gelatine block, rabbit's leg and pulmonary lesions, their motion was comparable in Tripoli and the British cities.[31] The bombing by aerial power appeared to target and affect a people not directly, but indirectly and at a distance. Its force gathered momentum along an environment or circumstance that propagated through the aerial subject.

Systems, Circulations and Ecological Warfare

The research of Zuckerman and Bernal papers over a disjuncture between the material and the social. It is clear that they considered material objects, physical infrastructures and urban mobilities to be ineluctably tied to collective feeling. Yet exactly how was explained away by physicalist metaphors of blast, explosions, cavitation, damage and disruption, and abstracted into a calculus of quantitative predictions and forecasting that was applied with 'too simple arithmetic' (Calder 1999: 175) to justify the British strategy of area bombing. Or it was melded into the simple *belief* that military figures such as Hugh Trenchard and later Arthur Harris held in the bomber. Bombing was "intuitive, a matter of faith' (Slessor in Verrier 1968: 37).

In other contexts the environmental milieu in which a population is targeted and reached has been made much more explicit (see M. Davis 2002). This is especially evident in the use of airpower as a force for counter-insurgent warfare – now an important blueprint for contemporary infrastructural and guerrilla violence. In this context Martin Coward (2006) argues that the complex ecologies of networked subjectivities are particularly vulnerable.

For influential airpower theorist John Warden (1995), the enemy can be understood as a system. Resembling the 'knock-out blow' envisioned by J. F. C. Fuller and other British proponents, if elements of that system could be destroyed the whole body-politic would fall apart into 'strategic paralysis'. Specific infrastructures tie the enemy state together in Warden's system, allowing it to operate as a 'single organism', 'entity' or network. Roads, airfields and factories all fall within the umbrella of critical infrastructure (Felker 1998) – the keystone for Stephen Graham (2006) of today's infrastructural warfare. And yet, the infrastructures we will examine are a little bit different. This section explores the aerial address of ecological infrastructures that support more fundamental and vital registers of life, forms of life that are 'biologically bound to the materiality within which they live' (Foucault 2007: 21; see also Coward 2006).

The Malayan Emergency of 1948–1960 is relatively unknown in relation to the major air wars of the Second World War, Vietnam and the Persian Gulf. Known as Britain's cold war (Deery 2007), the Malayan crisis has inspired political commentators, historians of airpower and military strategists as *the* classic experience of colonial counter-insurgency (Kormer 1972). British experiences in Malaya were later drawn upon to inform the American strategy in Vietnam during the 1960s and 1970s and the emergence of 'effects-based' psychological operations discussed in the following section of the chapter.

The Emergency was ignited following the British reoccupation of Malaya after the Japanese surrender in 1945. Malaya was a valuable colonial asset to the British, who had interests in rubber plantations and tin mines. An 'embodiment of the ideal colony' (Brockway 2002) with its imported plantation workers, anti-malaria medicines and species of rubber, Malaya is a fascinating model of colonial administration and management. Malayan European plantation owners employed Tamils to oversee Indian and Chinese migrant workers who carried out the manual labour involved in tending to the plantations. Taking fuller advantage of the workers, as Richard Baxstrom (2000) suggests, the European owners performed their own manner of biopower upon the migrant Chinese and Indian populations. Their power to take 'things, time, bodies and ultimately life itself from their workers' (Foucault in Baxstrom 2000: 50) rested upon the assured prosperity of the plantations as a source of *production*. Overseeing a climate of terror, formal regulation including access to work quarters, the amount of time that constituted a workday, living arrangements, sexual relations, and all manner of indicators of the worker's life, the British colonists were able to cultivate the migrant population into productive subjects.

The operation of British colonial capitalism in Malaya reached crisis point as tense labour relations with the migrant workers of profitable tin mines and rubber plantations broke out into widespread insurrection and the assassination of plantation managers. The 'squatting' Chinese population – who made up the Malayan Races Liberation Army (MRLA) Communist Terrorists (CTs) – were believed responsible for carrying out the terror. The reaction by the colonial government was severe. Pushed into the dense jungle environment of the region, the 'deep cover' became the habitat in which the MRLA and executives could rest, retrain and launch their operations. The Emergency, which gave rise to the term 'hearts and minds' (Coates 1992; Nagl 2005; O'Ballance 1966; Stubbs 1989), saw the involvement of psych-warfare employed by the Operational Research department of the British War Office to win over or convert terrorists. The Emergency also originated the marriage of the calculative and bureaucratic technologies of statistics and pre-diction which would be later seen in Vietnam (Gibson 2000), ethnological and biological expertise of jungle eco-systems, along with the eye of colonial administration and power. From the gaze of Operational Research, Malaya became a laboratory, 'abstracted into the universal space of positive science' (Gibson 2000: 81). For those on the ground the experience was quite differ-ent: Malaya was physically visceral, but it was also atmospherically debilitat-ing in its heat and humidity (Barber 2004; Follows 1999).

At first the Emergency took the form of a game of cat and mouse. The RAF became heavily involved in the crisis as they were called upon to iden-tify the location of CT clearings in the jungle canopy. The imprint of inhab-itation patterns upon the jungle milieu made the operatives' presence and activities visible from the air. The thick jungle vegetation was randomly dap-pled with open-air plots (known as *ladangs*) which had been cultivated in order to grow food crops for the terrorists to survive. The British control over the circuits of food supply and the local economy of trade and exchange meant the terrorists had to survive almost entirely off the jungle land. At first the cultivated clearances were one of the only signs of the CTs' where-abouts.[32] Reports show how between January and August 1953. 46 elimina-tions had occurred in what were termed 'killing grounds', open cultivated areas that acted to 'bait' CT operatives so as to trace them.[33] Like the Brazilian SIVAM/CENSIPAM system of aerial reconnaissance, the aero-plane became a platform for viewing as pilots were directed to identify not the operatives themselves, but their impact upon the surrounding ecology. Aerial photos and air surveys enabled intelligence officers to plot the loca-tion of terrorist hide-outs and settlements.

It was precisely the *bios* (Saint-Amour 2003) of the terrorist's 'dynamic habitat' that the British wanted to target. Pilots could determine food dumps, cultivated areas and settlement camps. By analysing the type of

Table 6.1 The probability of projectiles needed versus their likely success

Probability	Chance of success	Projectiles needed
0.01	1/100	70
0.05	1/20	340
0.10	1/10	690
0.5	Even chance	5,560
0.95	Near certainty	74,000

Source: D. F. Bayly-Pike, 'The probabilities of success in area bombardment', Operational Research Unit, Far East, MEMO no. 2/57, 1957, AIR 291/1701, The National Archives.

vegetation visible as well as the extent of its growth, along with the patternings of clearings either cut by tree felling or burnt, officers could highlight the physical construction of settlements. Plotted over the irregular jungle vegetation, they could build up a grid of possible locations to be targeted. In this manner, the effects of particular attacks upon the cultivations could be elucidated by further photo surveys.

The jungle was more than an indicator revealing the CTs' activities, for to make visible was a technique of the terrorists' own. CT environmental interactions were construed as 'militarily a method of tying down and harassing the enemy', frustrating the British and sapping their morale (Lewy 1978). Operatives were instructed to 'clear and plant', mobilizing the masses of the jungle to frustrate badly the British and Malayan forces: 'It would be very arduous for its soldiers; they would find the strain unbearable and their moral[e] would be thus gravely affected.'[34] The message instructed the CTs to strengthen their links with the local Malayan population, to 'propagandise and agitate among them and call upon them to free and return to the jungles to cultivate'.[35] With the intention of producing 'frustration' and 'low morale' for the colonial administration, the rhetoric blended the outcome of agricultural output with positive success. Clearing and planting could 'elevate the productive positiveness of comrades and anticipate the reaping of more beautiful harvests'.[36]

British bombing operations drew upon the experiences of the Second World War to calculate bomb effectiveness and required payloads and frequencies for either lethality or the benefits of *harassing* the terrorists. The futility of area bombing was revealed when one Operational Research report examined the number of projectiles that would be required to bomb six men hiding out within 4 square miles of forest. Measuring success by the falling of one projectile within 10 yards of at least one CT member, the grid in Table 6.1 was produced, tying the probability of success with the number of projectiles needed.

The direct effect of these bombs was not their greatest asset; indeed reports admitted that they did 'not know how frequently or for how long a period it is necessary to go on bursting bombs or shells at that unknown distance in order to produce a marked and lasting effect'.[37] One report suggested looking back to 'our own memories of mortaring and shelling and consider the effect on our own morale'.[38] Comparisons were made problematic by the fact that in London bombs that missed their targets would invariably have produced causalities among other neighbours, communities and colleagues, whilst 'those which miss the CT merely blow down a number of trees or make a big hole in the swamp'.[39] The material habitats of the city and the jungle were not immediately comparable (Coward 2006).

The very make-up of the Malayan CTs posed further problems for comparison with the Second World War. It was suggested that 'there are probably basic differences between the psychological make-up of the average Londoner and the average Chinese CT'.[40] Air operations began to be premised upon a science of speciation, developing what Anderson (2010b) has termed 'environmental'-type interventions which were aimed at controlling and mitigating 'a complex environment of insurgent formation'.

Speciation

The Malayan Emergency was consistently viewed through the colonial lens of racial division and suspicion. In films and images of the Emergency, distrust was interpreted through the surface of the face, an indicator that could speciate a civilian from a terrorist. As Carruthers demonstrates, the implication of this imagery is that 'one can tell a terrorist from his appearance, and part of this evil consists of being simply Chinese' (Carruthers 2005: 81).

Through reconnaissance missions and the interrogation of surrendered terrorists, the British and Malayan forces embarked upon considerable research in order to understand the biological and psychological characteristics of the terrorists in relation to the jungle habitat. As exemplified in films such as *The Planter's Wife* (1952), the rubber trees were presented as a cherished object requiring protection. In the film the jungle came to stand for the protagonist's marriage; with images of terrorists slashing, extorting food and shooting traitors, the colonist's place as paternal protector was fixed. In this context, the British asked: could CT groups sustain their activities whilst 'wholly dependent on wild jungle produce'? What would the 'effect' of the destruction of the cultivations have on the CTs' economy, their ability to continue, and how, once this knowledge was synthesized, could they be induced into surrendering?

Using the techniques of aerial reconnaissance resembling those discussed in Chapter Four, detailed studies of the trees and the heavy jungle cover were carried out. The variable depth and width of the forest gave the effect of a 'perforated canopy' about 50 foot thick and some 50–100 foot above ground level.[41] The flora created a sub-canopy world that proved an obstacle to detection from the penetrating gaze of aircraft above, whilst it permitted a ground-level free-flow of mobility and activity. The jungle insulated the CTs from the British and Malayan vision and their reach. Voice aircraft projections, bomb and shell blasts all had to be tested in order to understand the jungle's interference with the colonial reach.[42]

Reports were compiled using interviews with scientists and observations on the habits and requirements of the local populations. Interrogations were conducted on captured or surrendered operatives in order to investigate how the terrorists lived in the jungle, focusing on means of farming, common foods eaten and their quantity. Statistical methods sifted through interview transcripts and pinpointed the occurrence of key words. By identifying the most commonly gathered foodstuffs, sweet potato was registered nine times with regard to crops, nibong eight times in relation to wild plants, and fish some 58 times with regard to wild animals.

Understanding the human population was particularly complex given the mixed ethnic make-up of Malaya – a consequence of colonial settlement as well as worker migration necessary to man the rubber plantations. Malayan aborigines were seen as broadly divisible into several groups such as the Negritos, the Semai and the Proto-Malay. The Negritos were described as a 'distinct race', whilst in contrast the Proto-Malay were a 'synthetic group often freely inter-breeding with other Aborigines, Malays and Chinese'. Whilst 'inter-breeding' might create problems for the British taxonomy of the Malayan population, how the population's ethnic groups related and interacted with one another posed further difficulties. Finding a 'certain degree of ecologic association' between these groups or 'human communities', the Chinese CTs were rendered as outsiders and parasitical: 'aborigines often supporting or working with the CT or being parasitized by them'.[43] For the British, the aboriginal peoples became an ecological infrastructure for the CT groups, making decisive action upon the CTs harder to accomplish without collateral damage.

Reports attempted to discover the basic and most likely needs of someone living in the jungle by analysing basal metabolic rates – what the body would need to survive and sustain its metabolism. From these calculations it was possible to identify what sort of food would be required to fulfil those needs and the sort of jungle habitat that would support the growth of that food. A typical paragraph from one report stated: 'An active operational CT might require 3000 calories per day but could exist, doing nothing other than collect and prepare food, with 2000 calories per day.' This sort of calorie intake could be sustained by 2–3 lb of fresh tapioca tuber and twice those rates in bamboo shoots and nibong. Measured in terms of weight and therefore space per year, this translated into 1,917 lb of nibong or therefore 0.2 acres of tapioca per year. What the report describes as a 'quiescent' CT would need

0.13 acres. The protein intake of the CTs was believed to come from anabolizing fats and protein from their large intake of 'carbohydrates along with hunting fish and game from the jungle'.[44] Given the possibility of a low-salt intake, reports postured whether the CT might suffer occasional dehydration and would be unable to sustain an efficient metabolic turnover rate.

In order to maintain an adequate enough diet, the report then looked at what different kinds of farming and food-gathering methods could provide. Looking at cycles of farming practices on *ladangs*, the intelligence saw some areas cleared by cutting, some burnt, and others managed by a system of crop rotation. Dry padi was planted before the tapioca, among which capsicum chillies would be sown, some crops being left in the ground and others taken as required or cropped when they were ready. The cycle of clearance and planting was found to have been carried out annually. Examining the food intake of a particular village, the report's authors saw that around 0.2 acres of tapioca, 8.5 pigs and 12 tahils of salt (1.2 lb) were consumed by an adult each year. The salt was supplemented by a fort shop nearby; the rest was grown by the CTs themselves. From the individual body of the CT the reports authors could extrapolate to the group and spatialize the CTs' needs upon the jungle environment.

As a habitat or environment for the terrorists' survival, the jungle became more than simply an indicator of CT activities. Through the investigation of the jungle's support to their lives, it became a target to destroy or deny.

Providing and denying life

Once this complex human and environmental ecology was more fully understood, a more forceful internalization of this environment was possible. As Jane Bennett might put it, containing an active 'power to exert forces and create effects' (2004: 133), food, trees, flora and fauna took on what geographer Jamie Lorimer (2007) could label a 'non-human' charisma, a force upon the vital capacities of the terrorists to live and continue in their resistance. Fearful that the CT influence on the ethnic and local Chinese Malayan population might tip into more widespread insurrection, every effort was made to divide the local from the influence of the communists.

Malaya's discourse of counter-insurgency entered from the vocabulary of biological sanitation. Resonance with the contemporary cleansing of Baghdad and Afghanistan are obvious (Gregory 2004). The squatters, 'parasitized' by the communists, would be dealt with as an exterminator might treat an infestation. General Harold Briggs stated: 'You can kill mosquitoes in your house with a spray gun, but they'll soon be replaced by reinforcements if you don't sterilize the stagnant water pools where they breed' (cited in Sioh 2004: 735). The communists were "'flushed' out so that they could be captured by the police cordon around the patch of forest being bombarded". In other operations, the process of 'delousing' an area took place with a combination of light machine guns, as well as light and heavy armoured vehicles (Sioh 2004). Gas drops were intended to 'seal' off the circulation of terrorists from their own tactics of dispersal, although testing many of the psych-war strategies proved awkward given the difficulty of the technical 'captivation' of British or Malayan Army test subjects using gas or propaganda voice broadcasts and leaflets, reducing their status to one of experimental subject (Agamben 2004).

Another geometry of 'removal' was underway as the British sought to take on the role of 'protector' to the Chinese and Mayalan civilians. In a massive reorganization, some 500,000 Malayan ethnic Chinese people were resettled into planned settlements under the Briggs Plan of 1950. Compressing the Chinese population to isolated villages, the British constructed an organized infrastructure of the movement of food supplies, the provision of education and health services, and mobility infrastructures of roads and streets, providing the platform for the perception of a better and more secure life. Retaining the 'good will of the people' could be achieved by improving the 'social standards and conditions of the lives of those people' alongside other social reforms.[45] As with Vietnam, winning the hearts and minds of the people was carried out by building new synthetic milieus (Gibson 2000).

In providing an infrastructure for one part of the population, the provision of the means to survive and live, if not live better, were subsequently removed from the CTs. 'Food denial' was integral as all routes for food to leak into CT hands were sealed up through the regulation and control of food circulation within, to and between the villages. Within the villages the cooking of rice was centrally controlled and distributed, ration cards were handed out, the gates to the villages were strictly monitored, and mobile 'food-check teams surprised people' with spot inspections.

Communist jungle-clearing crops were destroyed by aircraft which sprayed defoliant chemicals – a precursor to the spreading of Agent Orange during the Vietnam War. Labelled in Parliament a 'vicarious punishment contrary to the principles of natural justice', and a 'technique of direct terrorism both immoral and ineffective', defoliants such as trioxine were combined with STCA, having effects upon three- to four-month-old sweet potato and tapioca crops, which began wilting within seven hours. The defoliants had killed all the sweet potatoes within five days.[46] At the same time as trials were made on food crops, tests were made on the jungle itself in order to discover whether the cover of vegetation could be thinned out sufficiently. In order to seek now visible terrorists, to serve as a target marker for potential strike forces, and to clear roadsides and areas around the settlements and jungle forts, defoliation meant avoiding ambush and the provision of cover for communists to hide.

By interrogating surrendered CT members, crop destruction was found to have been effective in causing 'considerable distress' from the temporary food shortage. Indeed, lack of food was one of the biggest causes of surrendering behaviour as the losses of destruction were rarely made up from other CT units. Only very rarely were attacked cultivations re-cultivated.[47] Reports postulated whether the CTs would suffer occasional dehydration, and would be unable to sustain an efficient metabolism. 'Lassitude', 'body wastage', 'vitality', the emergence of low morale and enthusiasm to fight were valuable outcomes of crop destruction.

In biopolitical terms the strategy of defoliation in Malaya denied only certain kinds of life (Westing 1983), performing what has become known as a brand of environmental terrorism (Schwartz 1998). Defoliation in Malaya was deliberately aimed at CT crops and jungle foliage in order to avoid the general population. As in Vietnam, the process went hand in hand with the semi-urbanization of the civilian Malaya populations into government-run villages (Galison 2001). A by-product was the over-accumulation and stratification of life, both urban and flora. Whilst in Vietnam the defoliants overstimulated plant life so that it actually exploded, creating a rotting and cancerous landscape of vegetation and human (see also Sebald 2003), in Malaya crop and settlement diversity gave way to jungle mono-culture and planned villages (Stoler 2008).

The British were anxious that they might send the communists *too* native and into total subsistence with the jungle habitat. Encouraging the enemy to regress in this way would see them 'adopting a more primitive, but not necessarily less efficient, food economy'.[48] Even though reports postulated whether this kind of economy would reduce CT 'fighting efficiency' because of the psychological uncertainties associated with more subsistence forms of a living, splitting the CTs into smaller guerrilla-like bands would mean they could become 'almost impossible to locate and destroy' and would have 'no necessity to try to hold a particular site or area against attack'.[49]

Adaptation

There appeared to be little doubt that the 'deep-jungle CT', in adopting an aboriginal type of food economy and rejecting the practices of manuring and crop rotation, was 'rapidly adapting himself to the jungle habitat'. Many a CT could be described as a 'near natural forest dweller, adapted to his environment and able to exploit its resources'.[50] However, the British could adapt too and absorb the techniques of the communist other – whose grounded intuitive and experiential knowledges were counterposed to the logic of Operational Research (Gibson 2000). Dropping down from the position of the aeroplane, the British sought the expertise of aboriginal trackers who were employed for their indigenous knowledge of the jungle in wayfinding and searching out CT operatives.

Marking a 'subalterns war' (Mansfield 2002), ferret teams were used which included Dyak headhunters to find and flush out CT camps and hideaways. The colonial forces had to think like a CT in order, for example, to predict the likely whereabouts and disposition of a CT camp. Points were to be considered which would 'subconsciously' 'affect selection of the site'.[51] Environments that could provide security were likely terrorist hideouts. Security was manifested in the protection from surprise 'by siting camps where movement is highly canalised, and any movement off these routes results in noise, or by siting behind a belt of open country, any movement being readily seen'.[52] From this knowledge, 'forecasts' could be made by going through a sequence of checks of the local environment.

As colonist led by the Oriental other, the British were taken down from the gaze of the aeroplane and the rational processes of Operational Research, to grounded and more embodied ways of knowing. Like the RAF in Iraq (Satia 2006), they learnt to appreciate the perceptual experiences of sounds, smells, the turn of a leaf, and the flattening of grass. Experiments with the Dyaks revealed their capacities of observation: 'a keen appreciation for the unnatural lie of plants, of disturbances in the natural settling of stream mud and of imprints left on wet earth. Time is judged by the fading of dead leaves, the exudations from bruised twigs or the squirming of an ant recently trodden on'.[53] Tests were even embarked upon to investigate the possibility that trekkers could perceptively 'feel' the presence of enemy operatives and the whereabouts of their camps: 'The reason for success that the trekkers "feel" the presence of another human being is probably analogous to the commonly felt "presence" of someone else on entering a dark and empty room.'[54] Although these techniques were tested and analysed by Operational Research, they were often found wanting: 'The unreliability of this "feeling" is demonstrated by numerical results,' concluded an OR report.

Air Conditioning

Here in Gaza, the Gazans have not slept for the noise of the helicopters, of the Israeli drones and the artillery shelling. These things were continuous all night and until now.

I am now speaking with you while walking on the street and, in order to hear you, I am putting my finger in the other ear in order not to hear the noise of the drones and the helicopters in the skies of Gaza.

This is the first time that we, in a week, slept and Al Jazeera tonight, today woke me up. I am hearing the drones still hovering in the skies of Gaza, continuously of course, to remind us that the Israelis are saying: 'We are still here.'

(Dalloul 2009)

These reports come from two Palestinian civilians describing their experiences to Al-Jazeera in January 2009 from the Gaza Strip. They describe sleep deprivation, continual terror and the struggle to continue ordinary lives. To blame are the sonic landscapes created by the Israeli incursions into Gaza. Even in the middle of the unilateral ceasefire, the noises of the shells, helicopters and the Unmanned Aerial Vehicles (UAV) (drones), or *zananas* (mosquito) as they have been named, reinforce the Israeli Defence Force (IDF) presence (Gaza.blogspot 2006). Through sound propagated from above and far away, the IDF were able to convey the message 'We are still here.' In the aftermath of the violence and arguable war crimes conducted upon the population in 2009, the whirring of the drones, their 'very distinct noise', cannot be separated from the terror conducted; they cannot be disassociated 'from the three weeks of precise bombardment and death which accompanied it' (Bartlett 2009).

In one main respect the UAV drones symbolize the Israeli occupation of the Gaza strip, which it now conducts 'indirectly' through a layer of vertical geometries or territories (Weizman 2002). For Marc Garlasco (2009), consultant to Human Rights Watch, the drones demonstrate how 'even without permanent garrisons, Israel continues to control Gaza's economy and infrastructure', ranging from its borders and airspace to its energy and monetary networks. And just as Israel might 'dictate from a safe distance what Gazans can eat, whether they can turn on their lights and what kinds of medical treatment are available to them', so drones give Israel the ability to carry out targeted attacks without having to risk 'boots on the ground' (Garlasco 2009).

This is one significance we might take from the drones' presence. The second is that their use actually performs some very old ways of thinking about airpower (Williams 2007, 2009). Now we are back with what Satia (2008) has recently called the 'uncanny echo' to the birth of the RAF's counter-insurgency policy in the desert. This was the formulation of warfare without territory. This was action and surveillance at a distance which disposed with the need to hold onto land or an army as the British terrorized Arab villages and settlements with the continual threat of aerial bombardment. The platform of the drones not only provides all the means for the delivery of death, but it also hoovers up the 'full spectrum' of activities on the ground below. Furthermore, the drones' simple presence marks a continual and age-old trend to attempt to reduce the material damage of air violence through its performance as a psychological tool. As for the British colonial policy, 'officials proclaimed that it worked more through the threat of bombardment than actual attack, gamely embracing "terror" as its main tactical principle' (Satia 2009). In this final section we explore how aerial environments are carved out as ephemeral and affective atmospheres of fear and perpetual terror.

Psych-warfare, morale and Shock and Awe

In 1996 the RAND Corporation (Hosmer 1996) published an extensive and somewhat tortuous review of the psychological impact of US airpower in the major wars of the twentieth century. Surveying research from the Second World War through Korea, Vietnam and the first Gulf War, the study considered the successes and failures of airpower deployed psychologically. Dwelling on the targeting on both civilian and enemy, the report made several recommendations for the waging of aerial wars in the future. A key element was a technique that captured and encompassed the wider strategy recommended: the art of 'air conditioning'. The right patterning of aerial assets, it was argued, could be used to 'condition' the enemy. This would cause armed forces to desert their equipment, to remain in their foxholes, and to be more likely to surrender. Conditioning would 'convince' the enemy forces of the following, captured in a scarily simple set of expressions:

> If you fly, you die
> If you fire, you die
> If you communicate, you die
> If you radiate, you die
> If you move with your vehicles, you die
> If you remain with your weapons, you die. (Hosmer 1996: xxxi)

Convincing the enemy of their death, expressed in this repugnant set of possible behaviours, meant a series of aeroplane-driven conditioning operations in order to generate specific perceptions, feelings and beliefs. By adopting operations that would maximize 'psychological effects' (Pape 1996), an air commander should attempt to concentrate airpower on targets around the clock in order to provoke demoralization; they should 'intimidate' by destroying food lines; 'shock' with devastating precision weapons deployed; and, by conducting attacks with near impunity, they would demonstrate that the enemy's 'resistance is futile'. In other words, the entire rhythm, volume and direction of airpower could and should be directed towards convincing the enemy that they cannot win.

The air conditioning recommended in the report is one among many components of what is known today as 'effects-based operations' (EBO) in military discourse (Bingham 2000; Carey and Read 2006), or 'targeting for effect' (Meilinger 1999). In order to achieve coercion and other 'axiological' political outcomes – from the surrender of the enemy to their withdrawal – a series of mechanical 'levers of power' translate military actions into a chain of effects that will eventually lead to those political outcomes (Meilinger 1999: 48; see Anderson 2010a). The demonstrations of airpower in the second Gulf War articulate a newly packaged yet familiar understanding of 'air conditioning' led by *Shock and Awe: Achieving Rapid Dominance*, the influential doctrine of Harlan K. Ullman and James P. Wade (1996; see also 1998). Embodying the central elements of 'effect-based targeting', Ullman and Wade state that military mechanisms and operations must construct perceptions which will lead to desired military and political effects. A series of effects are initiated in order to alter the enemy perceptions or *environment* which will conclude desired outcomes. For every action, they repeat, 'there is an equal, opposite reaction' (1998: 22).

Looking closely at *Shock and Awe*, these kinds of effects are *named* as the sorts of affect-states we are familiar with by now, and they are activated by applying pressure to *stressors* (Huss 1999). The 'shock' of 'Shock and Awe' refers to the ability to intimidate, 'to impose overwhelming fear, terror, vulnerability and the inevitability of destruction or defeat; and to create in the mind of the adversary impotence, panic, hopelessness, paralysis and the psychological incentives for capitulation' (Ullman and Wade 1998: 13). Awe is addressed through intangible measurables, not necessarily grasped, but, as Anderson (2010a) argues, verbalized as 'brilliance', 'competence' and 'irresistible'. Whilst an object such as morale is seen as a 'fuzzy subject', advocates of its targeting suggest that 'it requires both pin-point accuracy and understanding' (Ash 1999: 34). From the battlefield, Lambert (1995) identifies 14 levers or stressors, including 'Noise', 'Ignorance', 'Claustrophobia', 'Isolation', 'Helplessness' and particularly 'Fatigue'. As with psych-warfare in Malaya, Shock and Awe necessitates the knowledge and understanding of how the adversary ticks, what their 'thought processes' and 'value systems' are, and where their stressors are located.

By compositing the enemy state as a complex human organization, effects-based targeting considers the state's responsibility to look after its citizens (Wijninga and Szafranski 2000). A people's basic requirements might be fulfilled through territory or the security of an environment in which to live – familiar to us now given the colonial strategy in Malaya. The state's role must be to produce an 'environment in which the people can secure basic necessities'. Thus, threatening the environment in which those basic necessities are secured can differentiate and target – with increasing accuracy – a taxonomy of hinge points leading to the collapse of collective morale, will and the network (Anderson 2010a).

Drawing on human-ecological models of societal biological development (Figure 6.3), Abraham Maslow's (1959) and John Warden's (1995) conception of motivations and needs serves as a ubiquitous model in strategic

Targets: wealth, bank accounts, finances, confidentiality, hearing in some ranges, olfactory senses, taste

Targets: sense of beauty, normal sexual function, physical coordination, mobility

Targets: friends, allies, cronies, loyalty of children and relatives, love interests

Targets: sense of well-being, reliable cognitive function, orientation

Targets: food, water, clean air, rest, reproductive function, ability to eliminate waste, health, life

Figure 6.3 The evolution of Maslow's and Warden's systems. (Adapted from Winjinga and Szafranski 2000.)

airpower as well as counter-insurgency doctrine (Anderson 2010b). Maslow's hierarchical pyramid of needs has become a notable schema of knowledge that is even used to understand how the passenger-subject might respond to the airport terminal environment, as we saw in the previous chapter. In contemporary aerial war and security, the consumer and the securitized subject merge. They are 'understood to be the "targets" of two seemingly distinct contexts and practices' (Kaplan 2006b: 696). Recognizing the enemy as 'part rational and irrational [...] motivated by reasons of both passions and policy' (Meilinger 1999: 55), this kind of aerial strategy is about modulating desires and needs in order to induce a 'behaviour shift' (Meilinger 1999: 51–2).

We have therefore an evolution of the British in Malaya, who sought to know the full spectrum of the Chinese terrorists in relation to the jungle milieu. Focusing on the conditioning of human life let us take two examples, one from the Second World War and the other during Israel–Palestine conflict today. These orbit around the *stressor* of 'fatigue' modulated through the environmental milieu and the critical variable of noise.

Sound, sleep and collective torture

'Sleep we must', wrote Ritchie Calder (1940) in his investigation into the effects of the Blitz on London during the Second World War. When visiting the East End, Calder found a child acting out the movements and sounds of a Spitfire aircraft, careering round the playground and darting in and out of the classroom doors, arms outstretched and sputtering machine-gun-like noises.

Amazed that the child could so joyfully mimic an aircraft, given what he knew about a recent spate of bombings, Calder inquired further. A bomb had fallen within 7 yards of the family's Anderson shelter. 'Oh, him,' said the child's mother when the reporter explained that the bomb appeared not to have made any impression upon the young boy, 'he doesn't know a thing about it!' (Calder 1940: 252) It appeared that the boy was asleep at the back of the shelter when the explosion flung his body to the floor and failed to wake him.

Calder identifies sleep as a peculiar form of 'shelter', a 'protection or a healer' for the scars of the air raids; it was something 'we cannot afford to be without; something we must have in spite of the raids. It is more important to our morale than any "pep talks"' (1940: 252). For Calder, sleep appeared to be far greater protection than any material form of air-raid shelter or any kind of propaganda message. If sleep was a way to protect and rejuvenate, what effect would the nightly bombings have upon this medicine, Calder asked? Going to lengths to illustrate the pain of the aerial bombardments on the poorer areas of London, Calder identified lack of sleep – and its correlative effects of fatigue and demoralization – to be a vital sign – a barometer of the impact of the Blitz on Britain. Sleep and its shortage threatened the ability of the population to 'go on'. Given its common association with what geographer Paul Harrison (2010 in press) has shown to be an apriori 'body-in-action', fatigue threatened the efficiency and productivity of the workforce (Douthwaite and Leak 1940), and it denied their 'cheer' and 'spirit'.

Many people during the Blitz encountered the stresses and stains of war, which took their toil upon the body. Some people did 'cave in' and undergo what was known as a kind of 'apathy-retreat' in which an extreme desire for sleep was felt (Harrisson 1941). People embodied the psychological as well as the physical trauma of the raids. Hospitals recorded severe cases of neuroses, panic, increased frequencies of perforated ulcers, nightmares, hallucinations and other 'somatic' factors (Brown 1941; Crichton-Miller 1941; Stewart and Winser 1942; Wilson 1942). Although the dominant method of treatment was to reveal the 'mechanical' cause of the patient's trauma and to make them aware of their emotional injury, sleep and rest were also used as methods of recuperation (Wilson 1942). For Calder, a programme aimed at providing time or space for sleep was absolutely necessary for the well-being of the East End population.

Predecessor to the drones we saw earlier in the section is the IDF manufacture of 'sonic booms' – the shock-wave created when a jet breaks the sound barrier. Following the withdrawal of soldiers and settlers from the strip in 2005, the IDF launched nightly attacks of sonic booms over populated areas of Palestinian civilians. Whilst the settlers were still in Gaza, booms had been completely prohibited; upon withdrawal they were actively encouraged. Here we begin to see a direct connection between Zuckerman's studies of the geometry of the pressure wave and the intangible viscerality of a simulated air raid.

One teacher from the American International School described the feeling of awakening at 3.00 am to a sonic boom: 'The night sky seems to burst as if clawed apart. Windows and my teeth rattle. My heart pounds; my stomach tightens. Surely we are being assaulted by bombs and we will all die' (Kapka 2009). Claiming that they were a response to terrorist rocket attacks fired onto Israel, an IDF spokesman maintained that the booms were intended to act as a 'message to the terrorists'. Far better than an actual air raid, the message was 'non-fatal' and did 'not do long-term damage' (Macintyre 2005). Interpreted as 'mock air raids', the booms simulated the imminence of an attack or reprisal but without the violence and presence of missile attacks or helicopter gunships. Fear, in other words, was indiscriminately dished out by the booms. The boom, as one respondent replied, 'comes from above and, unlike artillery fire or a missile impact that can be located by the direction of the sound, it wraps itself around a person as if he is right in the middle' (cited in Karmi 2006). 'Look at how people react when there is a sonic boom,' psychologist Ahmad Abu Tawahina pointed out: 'Either they start laughing or they almost try to jump inside themselves' (cited in Karmi 2006).

The result of these effects upon the population as a whole, however, was far more serious than the IDF suggests. As a form of 'collective' punishment or torture, the acts were deemed illegal in the eyes of those who brought the case to the Israeli High Court. As one report continued, while it was true that 'liquidations and shelling are much worse', they suggested that 'the thinking behind the use of sonic booms is no less chilling: Israel has stuck with its old and bad policy that believes in "searing the consciousness" of the entire population so that it will pressure perpetrators of terrorist acts to stop their activity' (Levy 2005).

The emphasis of the dispute was directed towards the impact the 'demon attacks', or 'night terrors', would have on children residing in Gaza (Pilger 2006). The evidence brought by the Gaza Community Mental Health Programme and Physicians for Human Rights showed how the booms had behavioural implications (lack of confidence, withdrawal and hyper-vigilance), cognitive effects (lack of concentration, nightmares, violent fantasies, depression), somatic effects (muscle pains, stomach aches, hyperventilation) and the emotional effects of nervousness, anxiety and panic.[55] 'We tell parents', said one doctor, 'that when there is a sonic boom, the first thing they should do is gather their children in their arms to make them feel safe. But we know from talking to the children that they can feel that their parents are afraid. When parents are afraid, to children it means everyone is vulnerable' (cited in Karmi 2006). Indicating a cycle of violence produced by the 'mock raids', the raids foment resentment among young people, who are more likely to become tomorrow's terrorists.

The line drawn between sonic booms as an act of 'collective punishment' or torture is incredibly apt given the similarities between the IDF's distanciated forms of coercive torture, and much more direct contemporary approaches towards torture itself. Designed to induce a form of regression, the much publicized techniques of torture work upon the same basic 'homeostatic' faculties of 'fatigue', 'sleep' and 'pain' in order to produce the 'debility-dependence dread state' – or an emotional and motivational reaction of compliance. Once more, controlling and manipulating the 'subject's environment' is of paramount importance, says the US KUBARK counterintelligence interrogation manual, in order to 'to create unpleasant or intolerable situations, to disrupt patterns of time, space, and sensory perception' (cited in Benvenisti 2004). Like Maslow's adapted model of needs and motivations, disorientation, basic biological needs, and more sensory variables of sight, hearing and touch are legitimate targets. Plunging the subject into the strange, disrupting the subject's diet, sleep patterns and other fundamentals, the environment is modified to create irregular biological and homeostatic effects in order to produce a behavioural outcome.

Abu Tawahina, cited above, who is the senior clinical supervisor at the Gaza Community Mental Health Programme, said Israel was engaging in a conscious strategy to reduce Gazans to a 'state of learned helplessness', to reduce the population to a more vulnerable and helpless state. 'The Israelis are trying to make trauma overwhelming, and that happens when it is unpredictable, unavoidable and uncontrollable' (cited in Karmi 2006). Even the noise of the drones as they reveal their presence acts in the manner of the sonic booms; the anticipation of the drones' attacks simulates their potential violence.

Conclusion

Sitting in the middle of a continuum of airpower strategy that seems to have begun and ended in the Middle East, Malaya demonstrates the aeroplane's insistence on indirect effects: effects upon the material habitats of its targets in order to create effects upon the capacity of the enemy to live. And, moreover, through the address of human and natural ecological interdependencies, it conjures effects that augment and manage perception in order to generate psychological affects, such as frustration, despondency and a reduced sense of morale. In short, airpower has been about immersive *resonance*, the violent transformation and concatenations of materials, objects, flesh and affects. At the heart of this examination of aerial power is the coupling of material geographies altered and transfigured by the aeroplane with the militarized and 'inner space' of the body (Orr 2006).

We can see both of these tendencies coming together in the most recent developments of airpower which emphasize the importance of the infrastructural. These are infrastructures acting as the 'the sinews of society' (Graham 2006), life even more 'mediated by fixed, sunk' *affective* infrastructures surrounding aerial bodies with capacities to live (McCormack 2006). The result is a much more 'powerful political and military weapon'. As the targeting of the communists' jungle habitat in Malaya was essentially the identification of an ecological and biological system, today's aerial warfare is a similar identification of a point of articulation. In Gaza it is the background reverberation of sound which acts as the critical and 'perpetual conjunction, the perpetual intrication of a geographical, climatic and physical milieu' to either provide or deny an aerial life (Foucault 2007: 30).

Chapter Seven

Subjects under Siege

[A]ll the action of the disciplined individual must be punctuated and sustained by injunctions whose efficacity rests of brevity and clarity; the order does not need to be explained or formulated; it must trigger off the required behaviour and that is enough.

(Foucault 1977: 166)

Warning

Like the sound of a woman mourning the death of her child, the ululation began slowly, then rose in pitch to a crescendo. And then, as if she was pausing for breath, it diminished for one, two, three, four seconds, before rising again to its unnerving climax. (Mortimer 2005: 118)

The warning sirens, imaginatively described here in Gavin Mortimer's text on the London Blitz, were known by several names: Wailing Winnie, Wailing Willie or the Wobbler were some among many (*Lancet* 1940). Their ability to pitch in and out along with the rise and fall of volume composed the siren's amplitude and frequency of warning. With this distinctive tenor and lilt – discovered to be the most effective in ensuring extensive and intensive attention[1] – the siren alerts were felt like 'a series of great sound waves washing over the city', as American war correspondent Ernie Pile expressed during his stay in London (Pile in Gaskin 2005: 185). They could awaken, terrify and even appease.

Now transpose this alarm system with another. Take Brian Massumi's (2005a) analysis of the terror alert (Homeland Security Advisory System) installed in the United States in March 2002. The alert was colour-coded, designed to alert the public to the nation's current state of (in)security or 'threat conditions' by way of a governmental–media function. The alert system modulates and its palette is not only colour. As I write, the alert is currently set to yellow or 'elevated', which, according to the Department of Homeland Security instructions, advises making a 'response to the steps recommended at levels green and blue' whilst also making sure to: 'ensure disaster supply kit is stocked and ready'; 'check telephone numbers in family emergency'; 'develop alternate routes to/from work or school and practice them'; and, lastly, 'continue to be alert for suspicious activity and report it to authorities'.[2] Although the alert prescribes written instructions of preparation, like the siren, its qualities are more than verbal; its tinted agitations complying with a 'common vocabulary, context, and structure' that can disturb and 'prompt' all.[3] Like the siren's wail, the alert 'could raise [fear] a pitch, then lower it before it became too intense' (Massumi 2005a: 31). The population's attunement to this mechanism was as an activation – an agitation to action.

Now finally take the many airport evacuations and emergencies experienced since September 11. In Montreal Airport during May 2005 an alert was issued. A package had been found suspected to be the 'toxic substance' anthrax. The airport was evacuated and shut down. After the turmoil eventually the substance was found to be simply flour. But what the flour could have been – its potentiality as anthrax – performed a sort of 'affective time-slip' (Massumi 2005b). The alert's amplification ran away into a full-scale airport emergency, centred on the object not as it was (flour), but as it could be. States of fear and anxiety, common background feelings in the airport terminal, infuse the object 'regardless of what it actually is' (Massumi 2005b: 9). The power of this scenario is that the 'identity of the possible object determines the affective quality of the actual situation' (Massumi 2005b: 9).

All three alerts are inspired by a threat. The Blitz siren was the threat of the air attack – although quite what that attack would entail was anybody's guess. Gas, incendiary or explosive charge, the shadow of the bomber loomed with a fuzzy content. Massumi's discussion of contemporary terror alerts show that they are similarly formless. The threat is there but what shape it will take is simply a guess. All three alerts are cast from the mould of the aeroplane: the shadow of the bomber and two aeroplanes crashing into the towers of the World Trade Center. But in all three, '[t]he disaster that arrives and the disaster that may be about to arrive have equal powers" (Saint-Amour 2005: 131). It is the warning that subjects the aerial body to a siege of danger. As a way to be 'ready for it by directing where its effects will be felt' (Saint-Amour 2005: 131), debates over the alarm of the siren encapsulate the address towards the aerial threat in both old and new experiences of aerial emergency preparedness. Danger and disruption are activated at their every turn; the siren-alert functions as the motif of this readiness, haunting the patternation of mechanisms aimed at disturbing and ultimately preparing the aerial body.

Introduction

In Britain's build-up to the Second World War, debates over whether the public should be warned of an imminent air raid continued from the experience of the First (Hyde 2002; Miller and Bloch 1984; Sutherland and Canwell 2006). Just how the warning of attacks would take place was a difficult question for scientists testing the use of flare explosions (maroons) and ultimately sirens. The rationale behind the warnings was debated thus: how can precautions be set in motion upon the event of an air raid attack? The precautions taken could not be maintained during 'normal' times because of the inconvenience they would impose upon everyday life. 'The sole purpose of any system of warning', decided the Committee of Imperial Defence – set up in the inter-war period to discuss the country's preparations for war – 'is to enable precautions to be taken which cannot be maintained permanently without serious inconvenience or the interruption of essential activities,'[4] Focusing on the potential disruption, the 'object will be to give the warning at the latest moment which will allow time to carry them out and to cancel it as soon as the danger is no longer imminent'.[5] The signals would have to be manned by someone who could respond within 10 seconds to a telephone call. Speed and reaction times of response and recovery were essential to the efficacy of defence and the mitigation of 'surprise' (Kam 2004). In other words, the state of action an attack would provoke could not be made permanent; it needed to be managed and activated at the right time.

The Committee realized that in seeking to reach every person within a given vicinity, the warning should be 'expected to create the maximum possible disturbance of normal activities within that area', although, problematically, this would mean that 'the risk of the alarm spreading to adjoining areas which are not immediately threatened is proportionately increased'.[6] The organization of systems of warning became especially difficult in city boroughs where local authorities took control over siren warnings across small and dense spaces of jurisdiction. Sirens did not respect each authority's boundaries, causing several local councils to seek government arbitration. The efficacy of transmission of the warnings was, of course, vital to their success. The siren was a 'weapon whose power rivalled that of the airborne bomb' – its power lying not in its ability to wreak material violence, but in withholding the opportunity for a 'kinetic outlet', whereby it would 'disrupt industrial war efforts and shatter the citizenry's peace of mind' (Saint-Amour 2005: 140).

The siren's ability to agitate was tested. The earliest tests looked to the value of the siren for its capabilities of 'audibility at given distances', where '[n]o special attempt' had been 'made to find a form of sound which is distinctive or peculiar'. Later siren experiments in 1938 found that they could

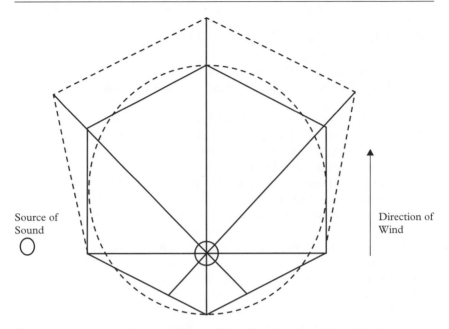

Source of
Sound

Direction of
Wind

Figure 7.1   Diagramming the siren's influence. (Adapted from Committee of Imperial Defence.)

transmit over 1 mile successfully, finding that the best range of pitch lay somewhere between 400 and 800 cycles, which would posses 'good penetrating power through the background of urban noises'.[7] Whilst the lower notes would penetrate further, the sirens needed to be adjusted to the 'human ear', which was 'more sensitive to the latter, and low notes are more likely to be obscured by background noises'. Environmental conditions played their role in the transmission of the warning. Patch fog, variations in air temperature, the direction of prevailing winds and the speed at which they travelled, these variables would impart their own effects on the range of siren audibility (see Figure 7.1). Informed by Dr Tucker of the RAF's experiments department,[8] the siren's propagation had to be understood in terms of its urban context, with buildings providing the most obvious obstacles to the sound's transmission. Sirens needed to be placed 'not lower than the general roof level in their neighbourhood', and they were not to be 'masked by a higher building close at hand'.[9]

Although earlier experiments had assumed that 'recognition of the signal would be secured (a) by the use of a standard code of blasts, (b) by the enhanced attentiveness which the risk of air attack would create, and (c) by the prohibition of the use of sirens and hooters for other than warning purposes', later tests determined that 'changes of pitch was more likely to attract attention than a signal of steady pitch'.[10] It was concluded that either sirens capable of the warble or a steady siren conducted with intermittent blasts would be appropriate. Both were to be of two minutes' duration, with the warning signal a fluctuating or warbling blast covering a frequency range of no less than 10 per cent.[11]

Local councils were directed by the Air Raid Precautions (ARP) Department to consider sirens.[12] Although the Home Office had tested many, they suggested that 'the vagaries of the travel of sound make it impossible to frame a full scheme without trial'.[13] Alongside the more general call of the siren, a distributed series of signalling was implemented. Police constables and air raid wardens were able to repeat and reinforce the public signals using whistles. For gas warning, hand rattles were used. For cancelling the gas warning and signalling the 'all clear', hand bells would be appropriate.

Securing and optimizing the efficacy of a warning increased its capacities of disruption. A careful balance had to be met by the government: 'The benefit to the community which is anticipated from a system of public warning must therefore be weighed against these disadvantages.'[14] If 'on the receipt of warning members of the public can take action which will contribute materially to diminishing the effects of an attack', then there was 'substantial argument in favour of their being given an opportunity of taking such action'.[15] The question became, 'What then can the public do? And how far will they be assisted to do it by any warning which it is possible to give them?'[16]

Through a policy dominated by an ideology of dispersal and evacuation, decision makers were concerned with the question of whether a warning might not only disrupt but bring the population into even greater danger than they would have been placed. Drawing upon the experiences of the First World War, reports found that 'public feeling was one of indignation combined with curiosity, rather than apprehension; and there was a real risk that a general notification that an attack might occur would have the effect of attracting into the open people who would otherwise have remained indoors'.[17] The spectacle of fires and destruction risked concentrating people into one space, which could be potentially catastrophic should a further bomb explode. However, the chances of such a 'misfortune' were considered really quite small, and the risks certainly less 'formidable' than if evacuation were suggested.[18] Planners envisaged the danger of civilians 'caught in the streets' and 'unable to disperse far, since traffic would be interrupted' along with 'undesirable congestion at railway and tramway termini', and concluded that the disruption would 'undoubtedly be a special object of attack if the enemy could identify them'.[19] Earlier developments of a siren warning system during the First World War had proved costly. When sirens sounded too long in Lowestoft, the furthest point east on the coast of North Suffolk, hostile aircraft were attracted to the area. Warning, in other words, could become a target in itself.

The main concern of a warning system was the disruption through the fear and potential panic it could incite. The experience had been to 'bring crowds of people into the streets and to cause considerable excitement and alarm', wrote the Chief Constable of Hull, noting the differential effects of sirens. There were those who were grateful for the siren: 'It has been opposed by a good many of the upper classes, but there is an almost unanimous desire for it among the other classes,' the Chief Constable testified. 'I have insisted on its being kept, having once started it.' On ordinary nights 'the people are perfectly "careless and secure"'.[20] Many people would evacuate and go trekking to the countryside in large numbers. Moreover, 'Sick people, old people, others who cannot leave their houses, and many of the better classes, who prefer to stay at home, are always greatly upset on "buzzer" nights.'[21]

The differential affect of the sirens – like the bombing – needed to be understood as certain groups and members of the population were more vulnerable to their wail. The sirens could be effective in soothing their populations, who would be alerted that they were being protected. For others, 'not the least good' could be gained by a public alarm: '[W]e think that it is liable to create panic, to disturb business and generally to disorganise the town.'[22] For Mass Observation, the sirens could begin to help people address their fears. Sirens could '*steel* people, both psychologically and physiologically preparing them for possible fear'.[23] The sirens were a form of inoculation. In its analogic urgencies, the siren contained 'some of the effects of actual bombardment in mild form'. Because '[u]nexpected fear has a much more shattering effect than expected fear', the siren, argued the Mass Observation reporter, 'keys people up for worse to come'.[24]

The issue became not merely how to warn, but how to prepare the public for a warning and how to behave on the event of a siren. 'This was a gradual process and depended rather upon experience than upon instruction,' advised the Committee of Imperial Defence. Certain sorts of action-response would have to be imprinted upon the population: '[T]he most that could be done was to inculcate certain general principles, the most important being that persons in the open should at once seek cover, in a substantial building if possible, but otherwise in the nearest available; and that those already indoors should stay there.'[25]

What these examples from the Blitz tell us is 'the function of the *alarm*'. The siren signals, it signs a threat in a manner so that everyone is activated into some kind of action. Threat, fear, emergency, all fold into one another as an event is *induced*, bringing 'the future into the present' (Massumi 2005b). Fears, atmospheres and behaviours are conjured and counterposed by efforts to manage and govern the aerial population. Sirens, in short, manufacture the emergency (Bishop and Phillips 2002b). Even tests of sirens had to be prepared for with warnings and notices, one in Birmingham stating it was being 'given in order to prevent any alarm being caused to the members of the public'.[26] However, people could be taught to deal with their fears and to prepare themselves for the event and the event of the alarm.

In this final chapter, we will see how the aerial subject is one who has been literally under siege from all angles – from efforts to destroy it and protect it. I suggest that efforts to defend and safeguard the aerial body have consistently addressed the subject not on its own but in relation to others – whose unity could either self-destruct through its 'capacity for revolutionary fervour or ecstatic, collective disobedience' or instead reinforce its security (Orr 2006). Throwing into relief contemporary logics of both aerial and non-aerial forms of securitization and, especially, preparedness (Collier and Lakoff 2008; Lakoff 2007), the chapter examines a history of systems of aerial protection aimed at securing the aerial body under siege from above and from itself.

The Anatomy of Panic

As we know, the science of the airport terminal is dominated by the figure of the PAX. In the study of evacuations, however, the abstraction shifts away from a shapeless number to physical models whose space is occupied by a mass of reverberating particles: objects who stand in for panicky passengers. The ways in which airports plan for the evacuation of panicked passengers have drawn on insights from fields such as pedestrian dynamics (Helbing et al. 2005; Waldau and Schrekenberg 2005). Couched in the analysis of complex systems, simulation models render passengers as mobile substances *en masse*, whose movements resemble the behaviour of the vortices of fluids (Helbing et al. 1997; Kerner 1999). The collective rush of passengers attempting to evacuate an airport on the agitation of a fire alarm is predicted, planned and quelled by the careful arrangement of architectural space that entrains passenger's emergent and unconscious movements. Orderable passenger queues and lanes coalesce as emergent behaviour from an apparently chaotic mass. Simple forms and design cues trigger the crowd's movement as a whole but from the bottom up. The traffic of the molar suddenly aligns. In the physical and psycho-social sheltering of the aerial body, addressed as 'logistical life' by an array of expert systems of knowledge, we see the development of a 'technological unconscious' of orderly and sequenced subjects (Thrift 2004b). We see how their protection is dependent, like the passenger crowd, upon one another.

'In wartime,' writes historian Travis Crosby on the history of evacuation from aerial threat, the government's 'prejudices made it easier to think of city crowds as an abstraction, as objects merely to be controlled and directed in times of emergency' (1986: 7). The secured position of the inter-war government was a belief that on the event of aerial warfare the population would react in a way that they could not control. During the war, scientists would develop typologies of *somatically* conditioned behaviours, as biological instincts superseded cultural norms of behaviour (Crichton-Miller 1941).

Their certainties, built upon the racial and class prejudices of biological degeneracy and the memory of the First World War, predicted an emergency born from the susceptibility of certain quarters of the population to crowd and panic. Jews were to be 'driven mad with fright', becoming 'a rather special problem'. The poorer working classes were 'likely to form a most unstable element – an element very susceptible to panic'. As Jackie Orr shows, both are associated 'with an array of pathologized others' (2006: 15).

As before, if the public were to be alerted of the onset of a raid, several questions became key: How to organize and prepare for this onset? How to ensure the population could be activated into doing something that would result not in disruption but in security? The idea of the panicked body, an irruptive individual capable of merging in and out of insurgent collectives – a threat to ordered society and a threat to themselves – frames these debates. Following Orr, these imaginations and experiences of panic became both the 'object as well as the medium of a power' (2006: 15).

The issue was bigger than that of the siren. Enwrapped in the discussion that engaged the British government in the build-up to and during the Second World War was the decision to act in advance of an air raid, well before the sirens had even gone off. In these deliberations we see resonance with contemporary discussions concerning the preparation of civil contingency plans. For Sir Charles Hipwood, a key figure in the organization of preparatory plans for civilian evacuation during Britain's inter-war years, the question was how to control a population under siege by aerial attack: 'In turning this question over in his mind he had first been struck by the hopelessness of letting things take their normal course. The effect of bombing on the totally unprepared population would be, he felt, simply devastating and would entirely demobilise London.'[27] If one agreed with this theory that it was essential to retain control of some kind, then one had to ask: how could it be done? Debating the question from the perspective of a panic, '[W]as it better to prevent people bolting or was it not? Was it going to be said that if people bolted it cannot be helped? On the other hand, if this solution was to be rejected, then some form of control would have to be instituted.'[28]

Hipwood's points were repeated in the House of Commons debating the Air Raid Precautions Bill in 1937. To act or not? To react or remain passive? The threat was positioned in Parliament as 'panic and a rupture' – the 'panic and the stoppage of national life'. The ending of life as they knew it came hand in hand with the eruption of panic. Both could be combated through organization on the ground.[29] In his speech for the Bill, Foreign Secretary Samuel Hoare stated:

> [T]here are some people in the country who say that it is no use attempting to take defensive measures, that air attack always breaks through and that nothing we can do will be of any effect. I do not take that view, and I am supported

by the whole course of British history. Time after time, the British Empire has been faced with great dangers, but never has it been faced with any danger and sat still, inert, inactive, despondent and despairing.[30]

The debate Hipwood and Hoare waged pivoted around the question of whether action should be taken and what action could help move the population out of areas of special threat. The question was a conditional one. If the population could not be relied upon to behave and move in an orderly way, then any actions that were taken would be ineffective. Hoare would emphasize, however, 'that unless the minds of the population were prepared, and the morale maintained, no form of control would be of any use at all'.[31] If the problem of a population 'bolting' could not be solved, 'it would be advisable to prevent any such spontaneous action, but he was at present at a complete loss to know how this could be done'.[32] A form of military disciplination (see *Lancet* 1938b) was needed, similar to Foucault's terms, wherein a signal could be reacted to immediately, 'according to a more or less artificial prearranged code' (Foucault 1977: 166).

The presumption of the debate was the object of panic, a lack of morale expressed through fear; aerial violence that escaped outwardly into bodily mobility. The assumption was partly articulated through the experiences of the First World War and several medical studies which were published in *The Lancet* (1917a, 1917b, Waller 1918). The government's salvation was the 'neglected' question of 'mass psychology, mob psychology'.[33]

The Committee of Imperial Defence had earlier brought together a forum in 1925 to consider this problem, drawing on the expertise of transport personnel to discuss the provision of mobility infrastructures and the events 'inevitable in case of any panic', such as 'big crowds' and the proportions of the 'extraordinarily calamitous'.[34] The Committee imagined that around 5 million people would need to be transported. For even the most 'stoical and unemotional population', mobility was going to be a necessity, requiring the authorities to see that 'it is not uncontrolled, that people do not flock to the railway termini in panic'. The rolling mass of the collective tumbling through space at speed was to be avoided with every effort: 'If such a [panic] rush takes place, there will be very serious difficulties in feeding and housing, as well as the possibility of casualties where the concentrations take place.' Even the issue of having employees to carry out the evacuations would be problematic: no bribe of an increased pension, not even a uniform for discipline could subdue the need to retreat from one's post.[35]

The Chairman of the Committee urged that 'such measures as are possible should be taken in advance, and that they should be very carefully thought out to prevent anything in the nature of a panic'. Panic would result if 'measures had not been taken, if they were uncontrolled and unforeseen'. The essence was to 'consider what it is possible to do to induce the public to face what is the inevitable with reasonable composure'.[36] Hipwood's solution was both psychological and physical preparation, rooted in the experience of industrial disputes (Bourke 2005).

> He was quite certain that the mind of the public must be prepared in advance. If the whole thing came as a surprise, then the situation would be uncontrollable. If, however, preparations could be made to control the morale of the population, then it would be necessary also to impose some kind of physical control. […] His impressions of the situation were that it would be essential to retain control, and in order that this control might be maintained it would be essential to prepare the minds of the people beforehand.[37]

Representatives on the Committee of Imperial Defence suggested various forms of military governorship, self-contained districts, martial law and container-like boundaries to protect the circulation of people in and out of London.[38]

How to 'induce the population to face the consequences of air attack' with spirit and positivity? Comparisons continued to be made with the threat and anxiety of industrial unrest.[39] For example, John Anderson, then Undersecretary of the Home Department, was not certain whether morale preparation could have the desired effect if it occurred before a specific plan was decided. They didn't want to 'agitate' or 'inspire' the population without having a definite plan to do so. Maurice Hankey 'said that there would always be the old difficult that the situation might begin with the most dramatic suddenness'.[40] Hipwood 'agreed and emphasised the necessity of having complete schemes prepared in advance'.[41]

The answer was three strategies that revolved around the "maintenance of morale'. Imaginary futures was one. Hipwood explained how 'he had been trying to visualise what would happen if the emergency was to occur next week. He felt that unless the population was prepared, we should be completely done in straight away.'[42] Some advocated a committee to 'examine carefully what could be done to prepare the mind of the population for the emergency which they envisaged', to be staffed by Lord John Reith of the BBC, Lord Beaverbrook, H. G. Wells and others, 'men used to influencing the minds of great masses of the population'.[43] Secondly, the social needed tying together – individuals sewn into a resilient public. And the final approach coalesced around the negative excesses of unlimited thought and imagination; praxis was needed. '[T]he worst thing possible would be to leave people with nothing to do, and that they would inevitably tend to get panicky,'[44] the Committee concluded. Anderson suggested that,

> If the population were set to work on definite tasks such as he had suggested, they would not only have something to occupy their minds, but would be doing extremely useful work. […] It was thought to be essential for the good

of the morale of the population that definite tasks should be set them, rather than that they should be allowed to stay at home and speculate on the immediate future.[45]

Prescribed sets of tasks, movements and actions were to be set.

Imaginations and Urgencies

Urgent and 'analogic' affects (Ngai 2004; Tomkins 1991) were composed through the looming shadow of danger. A manner of air-mindedness – a comprehension of being seen from above – emerged in these affective imaginations. Military scientists responded. Smoke screen tests began as a means to veil cities and vital areas, to undo their status as targets and make them 'indistinct or unrecognisable and therefore delay or defy identification'.[46] Pilot testers suggested that their 'attention is always attacked by smoke: advantage may be taken of this fact by using smoke as a decoy or to induce uncertainty by multiplying the target'.[47] Gas, as an ephemeral and seditious object evading both capture and comprehension, was the most anticipated source of 'frightfulness' from the air (Haldane 1925; Harris and Paxman 2002; Spaight 1939). Fearful of what Peter Sloterdijk (2009) terms atmosterror, in Britain authorities worried about the proclamations of radio broadcasters such as Professor Noel Baker, who transmitted the following dark future in 1927:

> 'In the first phase of the next war,' says a high authority, 'there is little doubt that the belligerents will resort to gas bomb attack on a vast scale. This form of attack upon great cities, such as London or Paris, might entail the loss of millions of lives in the course of a few hours. Gas clouds so formed would be heavier than air and would thus flow into the cellars and tubes in which the population had taken refuge. As the bombardment continued, the gas would thicken up until it flowed through the streets of the city in rivers. All gas experts are agreed that it would be impossible to devise means to protect the civil population from this form of attack.'[48]

Paul K. Saint-Amour shows how inter-war modernist authors who painted dystopic pictures of an urban demise acted to 'archive the city against the increasing likelihood of its erasure' (2005: 131–2). Today 'fantasy documents' (Clarke 1999) anticipate scary yet sometimes unlikely or, alternatively, far too 'reasonable' scenarios that work with their symbolic and affective capital to 'reassure' (Clarke 2006; McConnell and Drennan 2006). The literal 'switching off' (Graham 2005) of cities in contemporary infrastructural warfare finds its ultimate vision in the simulated consequences of an air war: 'Great cities murdered in a night; the centres of civilisation murdered and smashed; London, Paris, Berlin, a heap of smouldering ruins made hideous by a million dead.' Baker's imagining is cinematic in scope (Hell 2004; Pevner 2004) as he renders a scene which is impossibly disembodied, yet provocative of sickening affects: a paused city soaked with the tension of an incoming bomber; the 'gathering roar of a thousand engines, and the launching of the sudden murderous attack'; 'a hundred mighty conflagrations in every different district of the town, followed by the stupendous crash of high explosive'. Smashed to smithereens after every semblance of shelter was lost, 'when every street was full of shouting men and women and screaming children driven mad by panic fear, there would come the deadly floods of poison-gas'.[49] Located as a form of biopolitical attack (Sloterdijk 2009), unlike the explosive shell or incendiary bomb, gas was perceived to be directed 'principally against the morale of a country' because it was intended 'to impose itself upon human nature as to weaken the power of resistance'.

Although speeches like Baker's were monitored by a concerned British Home Office, the anticipatory futures he described were soon addressed in Parliament and promulgated by the Home Office with the hope of emulating Baker's urgencies. Their fictions 'rehearsed their readers for such an eventuality, drilling them toward readiness if not toward mastery of anxiety' (Saint-Amour 2005: 156). Hoare's speech to the House on the Air Raid Precautions Bill suggested that the 'greatest danger to civilisation' was the 'knock-out blow' against which Air Raid Precautions would be directed. Hoare went on: 'I urge upon the House the great urgency of this problem. I urge upon the country the great need for the fullest co-operation between all citizens.'[50] Hoare's stressing of the urgency of the problem is most revealing of the beseeching *imaginations* that government policy and especially Air Raid Precautions disturbed (see Kendall 1938; *Lancet* 1938a). People had to be told of the worst in order to prompt them into action.

For Baker and those critical of ARP, the government were attempting to create a *machinery* focused on and, indeed, generative of threat – the bomber, panic – which would be answered by the precautionary measures they suggested. 'What the Government are trying to do is plain,' stated Baker. 'They are trying to create a feeling of confidence in the minds of the people before war comes.'[51] The book published by the Cambridge Scientists' Anti-War Group (1937) took a similar line, noticing the dominance of panic within the speeches and discourse of ARP officials. 'Seldom do they fail to mention that panic is the greatest danger which they fear' (1937: 13), the scientists remarked.

Dissent took several forms. Firstly, like the contemporary terror alerts, ARP was seen as a modulator of emotion (Massumi 2005a). General pronouncements 'seem to be designed at times to make our flesh creep, and at others to be intended as a dose of soothing syrup for our nerves,' noticed J. M. Spaight (1939: 50). For the Cambridge Scientists' Anti-War Group, ARP appeared to be a form of domination of the kind we associate with Foucauldian docility. Panic was something that would ebb and flow, after all:

'This is the panic which 'was to be dreaded more than anything else.' This is the panic which it is the main object of Air Raid Precautions to avoid. Air Raid Precautions are an essential part of war preparations; they are a part of those preparations in which the acquiescence of the people is vitally essential' (Cambridge Scientists' Anti-War Group 1937: 53). Others who noticed the tonality of fear in ARP suggested that mobilization might be the last thing the precautions agitated. 'The purpose is rather to bring home to the nation the need for urgent preparation to meet it. Yet the effect may be the dissemination of general despondency, of a feeling of hopelessness, of the precise mentality the reaction of which to shock-action in the air may be just what an enemy would desire' (Cambridge Scientists' Anti-War Group 1937: 51). 'It is no exaggeration,' wrote an Army Lieutenant-General to *The Times*,

> to say that hitherto those responsible for ARP from Cabinet Ministers to ARP wardens, have terrorized the civilian population by impressing on them that salvation lies mainly, if not wholly, in burrowing underground, the deeper the better. As one ARP official explained to me recently, 'If we don't put the fear of god into people they'll do nothing.'[52]

For others, Air Raid Precautions appeared to be as much about preparing for the psychological effects of an air raid as about actually preventing a raid from taking place. It was quite plain, for Noel Baker, 'that many people who advocate air-raid precautions do not really believe that they will give much defence'. Keeping the people quiet, employed to 'prevent panics' and to 'maintain the morale of the civil population', ARP would not be able to do anything 'serious to defend the civil population from death and wounds'.[53] It was noticed that a 'large part of the service provided in this is merely to clear up the mess after the raid has taken place, not to prevent lives being lost'. ARP dealt mostly with bringing people into decontamination centres, putting out fires and clearing up the streets.[54] Like contemporary preparedness ARP did not necessarily prevent, but rather it responded and reacted (Collier and Lakoff 2008).

Vigilance and the Social as Circuit

One response to the aerial subject under siege has been urgencies – pulling and provoking response through dark imaginations. Another is a sense of collective and personal responsibility for how such futures should unfold. As Elaine Scarry shows in her analysis of United Airlines Flight 93, which was hijacked on September 11, the best defence against aerial violence has been not according to a simple hierarchical structure of external military defence, but on the basis of a more egalitarian form of citizenry. When civilian passengers became 'citizen-soldiers' and acted to seize control of the aircraft away from the terrorists, Scarry writes, 'it was their own lives they were jeopardizing, their own lives over which they exercised authority and consent. On the twin bases of sentient knowledge and authorization, their collaborative work met the democratic standard of "informed consent"' (see also Fritzsche 1992).

Populations under aerial siege have not necessarily fulfilled the same kind of deliberative process of self-governance that Scarry reveals in the micro-spaces, times and pressures of the Flight 93 cabin; indeed the opposite is often implied. And yet the collectivity of everyday citizenship has been the bedrock for aerial self-defence. Contemporary activations of civil watchfulness see 'attention – policy, analytical, and public' orientated towards 'the problem of governing security or, more precisely, the circulation of insecurity' (Salter 2008a: 10) by coordinating populations into a fusion of vigilance. As millions of dollars are pumped by the DHS into the American Trucking Association 'Highway Watch' programme, truckers are encouraged to be vigilant. The events of September 11 haunt proceedings once more, as against the backdrop of the burning Twin Towers a promotional video 'represents truck drivers as first responders, displaying "I am Highway Watch" badges alongside the US flag on their uniforms. "All of those eyes out there on our nation's highways," proclaims the voiceover' (Amoore 2007: 216).

The United States' 'Be Ready' campaign marks a similar juxtaposition of the wreckage of the World Trade Center with advice to 'be aware of conspicuous or unusual behaviour', which should be 'promptly reported'.[55] In the surveillance prescribed by contemporary security discourse, we see examples of the everyday and expert gaze orientated towards foresight. Activated towards threat and emergency, community cohesion and collective responsibility appear unlikely bedfellows.

Take the writings of John Langdon Davies (1939), who used fictional narratives of the traditional conception of the family unit to impress the need for preparation and organization during the Second World War. In a scene inspired by *The Swiss Family Robinson*, Mr Jones compares his family's situation to that of being cast on a desert island:

'If we were all shipwrecked, how would the Jones family see it through? And when you come to think of it, an air raid would be pretty much like a shipwreck.'
   'Why Dad?' asked Mary, just home from the laundry where she worked.
   'Well,' said Mr Jones, 'if you want a drink of water, what do you do?'
   'Turn on the tap, of course,' said Mary.
   'Yes; and if you want some light, you switch it on; and if you want to cook, you turn on the gas.'
   [...]
   'If we were on a desert island we should have to be able to do all these things, and a thousand others, which somebody else does for us as it is; and exactly the same would be true if there was an air raid.' (Davies 1939: 4)

In this excerpt, the family realize that an air war will disrupt their most basic patterns and daily routines. Their father's lesson is that what they would normally do to keep on living would stop; it is going to be the family's responsibility to take care of themselves.

Jim, the son of the family and the first to understand his father's message, quickly sets his mother straight when she asserts, 'But it's the Government's duty to look after us in war time':

> 'No more than it would be the captain's job to look after us on the desert island, Ma,' said Jim, who had been listening to his father with both ears. 'Maybe the captain would have got drowned when the ship went down. I think Dad's right. I vote that the Jones family gets down to the job of preparing for an air raid, just as if we were on an adventure voyage and going to be shipwrecked on a desert island.' (Davies 1939: 4)

Davies tries to suggest that the family's preparations should be fun, an adventure. Before the family even gets hold of a *treasure map* of the local area – where the treasure consists of vital services and shelter – they try to 'imagine exactly what an air raid would be like, so as to have a clear idea of just what we ought to expect' (Davies 1939: 3). The structure of the familial unit found backing in the view of medical scientists, who proposed that an internal support could be developed through the organized structure of the *group*. Panic, one article suggested, 'is probably best allayed by doing something for those one loves' (Rickman 1938: 1293).

Today's emergency plans are enthusiastic about the same kind of imaginative role play. 'Talking it out', suggests the Ready.gov, a website of the US Federal Emergency Management Agency (FEMA), can be a good way to create a 'clear family emergency plan' that involves all members of the family. 'Where would we meet?', 'How would we remain in contact?', 'What would I do if I were at school?', 'What would we do about our pets?', are questions that gesture towards the problems faced by a family in an emergency and a means to solve them. The form of Davies' treasure map is used in 'Hidden Treasures', an interactive anthropomorphized family scene that diagnoses and trains the child's level of awareness in the possibilities of their surroundings (Figure 7.2). The child is encouraged to participate by finding several 'important' objects when planning for an emergency; these include a family 'supply kit', telephone and photos of relatives reminding them to know an 'out of town contact'. Building this kind of plan, it is suggested, can help 'establish a sense of control and to build confidence in children before a disaster'.[56]

Before the Blitz, the necessity for these levels of personal activation was proposed at all levels of public life,[57] although precautions could only take place in a local and 'civilian field'. '[T]he main work', declared Samuel Hoare, 'must be done in the individual localities and must be carried out by civilian

Figure 7.2   Ready.gov's 'Hidden Treasures' map. (FEMA.)

men and women.'[58] The family unit transposed onto the community was the best response. The cybernetic equivalence of body and machine discussed in Chapter Five added up to the notion of a public biological and familial nervous system. Panic, and its antidote, morale, would be fed through social conduits blurring the boundaries of individual and group, 'producing, transmitting and receiving information through intrapsychic or interpersonal processes' (Orr 2006: 106; see also Wiener and Bigelow 1943).

In Germany, preparations for aerial warfare were taken on the home front by a civil society forced 'to transform themselves into "soldiers" whether it was on the battlefield or on the home front that they could see action' (Rieger 2005: 252). The German Air Defence League and Civil Defence League both took Air Protection or *Luftschutz* seriously, requiring by 1935 all ablebodied citizens to take part in air raid protection (Davis and McGregor 2007). Certain models of behaviour would set standards whilst they would lead the population in their preparations and reactions to an air raid.

The air raid warden (Figure 7.3) was required to reinforce the collective, circuiting the social in order to tie people together through more formal networks of organization and coordination. If these pathways could be

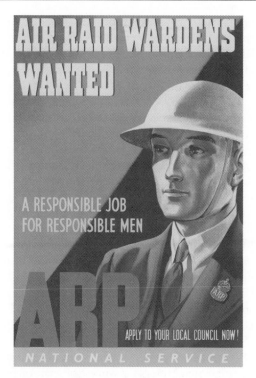

Figure 7.3   Recruiting the wardens. (Crown Copyright.)

forged, the population might not panic but be induced into a response. In Germany the Civil Defence League became an 'instrument of social control'. Its wardens conducted drilling and exercises and oversaw blackout procedures. If anybody tried to 'resist these intrusive measures', writes Bernhard Rieger, then they were 'liable to be charged with endangering the nation through negligence' (2005: 252).

In Britain, wardens were required to know their neighbourhoods, capture and learn formal and informal knowledge such as: 'who lives in his sector? Which of them can be relied upon to help in emergency? Or which are especially likely to need help?' They should know the location of fire hydrants, places of special danger, where the local respirator store is, what telephones are in the area, where public shelters will be; the location of ARP services in the area; names of doctors, chemists. The list went on.

Whilst they should be required to encourage people to 'make their own precautions in their homes as good as possible' (Air Raid Precautions 1938a), they were also key figures in the actual event of an air raid. Wardens needed not just to know, but to anticipate and act within neighbourhood networks of association. Wardens would assess neighbourhood vulnerabilities and head up control rooms to plot the damage of a raid before sending the information to the relevant emergency service for deployment.[59] The ARP officer's communications were routed through a schematic logistics of timing and sequence.

Wardens were required to direct and be helpful to people generally (Strachey 1941; Thomas 1939). It was made clear that while the warden was not a 'policeman, nor a special constable and, therefore, not able to exert constabulary powers', he could, 'by his example and his readiness to help, exercise a powerful influence in time of stress'.[60] According to the ARP memorandum, the warden could communicate several affects. He should have the 'confidence of his neighbours'. He should be able to 'calm them and reassure them, to help them with some immediate aid'. In short, the warden should 'do everything possible to avert panic' and stop people rushing out into the streets.[61] Generally acting like 'a good neighbour', the warden should be 'a responsible member of the public chosen to be a leader and adviser of his neighbours in a small area – a street or a small group of streets – in which he is known and respected'.[62]

Mimicry, the staple of contagious affectivity, was applied to the air raid warden's ammunition (Strachey 1941; Thomas 1939). In Birmingham, circulars recruiting officers declared that the ideal candidate should avert panic through setting 'an example by his or her own calm behaviour'.[63] Wardens would 'shepherd' citizens to shelter and, by demonstrating 'coolness and steadiness among their neighbours', reduce 'the risk of panic and loss of morale'.[64] In a visit to Birmingham, John Anderson, now Minister of Home Security, aligned the air raid officers with the stature of a super-citizen, a sufficiency of such 'who will turn out to face all risks and to give their help'. The complex and multiplying conflagrations could arrive with such speed and unpredictability that a networked and 'nucleus force of trained sentinels covering both day and night and every day of the week' was essential.[65]

If the morale of the population could be accessed by wardens through these pathways, then they were open to the population to sustain and reinforce one another. Morale was given physical shape as an instrumental tool to be gripped and wielded. With positive sentiment a transferable object, it became the population's duty to pass it on and to stop the movement of negative mood.[66]

Recalling her walk across London on her way to her office one night during the Blitz, journalist Ellen Harris describes meeting a man who had been taking cover in a shelter in Lincoln's Inn Fields. Almost in tears, he said, 'What are we going to do? We can't go on like this, we've got to seek peace, we've got to say we must have peace. We can't go on in this way.'[67] For Ellen, the man was evidently 'very panicky' and needed to be calmed and pacified should he *infect* anyone else with his disaffection. 'Do you realize that you're playing right into Hitler's hands?' she said. 'This is just what he's setting out to do. If he can do this to you to get you into this state and you come here and you start on me and I join in with you; and I go up the road and tell somebody and you do the same to somebody else, no.' Doing so would 'get people in the state of mind and their morale goes'. For Ellen, the answer was to rise to the level of Scarry's (2002) citizen-soldier; she urged the man to remember that he was on the 'front line' as if 'in the trenches in the last war. [...] This is what I'm telling myself all the time, this is my war effort. And this is your war effort. [...] And he said, "Thank you, thank you very much." And he was bucked up and he went off. I hope I did him some good.'[68]

Continuing her journey, Ellen's walk took her past numerous fires that were spreading their way through the night-landscape. Coming across an old Victorian house which had been turned inside out by the flames, she found herself unmoved by mothers scuttling around with their children's prams carrying babies and prized possessions, yet particularly *moved* by a burnt birdcage, 'still hanging in the window and a little, dead canary'. 'Now can you understand that?' she asked. 'I just burst into tears going up the road. I thought, "This is terrible."' Meeting a butcher providing free food to local people and firemen who were 'flung' across the front of their motors through exhaustion, Ellen's morale appeared to be seeping away. '[J]ust for once on that day, I think mine was going,' she recalled and went on:

> I got back to the office and I said, 'Oh, God its terrible.' And I burst into tears, you see and the editor said, 'Now, what's the matter? What is it?' I said, 'I've just been out and the firemen are still fighting the fires and half the firemen', I said, 'are dead on their feet. They're flinging themselves down across the engines to get ten minutes' sleep. You can't get anything, anywhere to eat, you can't get a drop of water or anything outside, this is terrible.' You know, I was looking ahead a bit, if this is now the beginning of it.[69]

At the depth of her despair and anticipating the worst, Ellen found that it was then the turn of her editor to remodulate his part of the circuit, 'it was his turn to push morale into me'. The editor told Ellen,

> '[Y]ou, of all people, I'm ashamed of you. You pull yourself together,' he said, 'we're looking to the likes of you to keep everybody going. Now,' he said, 'there's a lot of youngsters around here and they rely on you. Don't let them see you upset, they're going to be upset. So,' he said, 'keep going'. And never once after that did I falter. But it was that day, all the fires going around.[70]

ARP wardens and staff were given lectures by the Mental Health Emergency Committee so they would be aware of possible mental health problems in time of crises.[71] Creating a 'smooth running machinery' and supported by a series of informal assistants to whom they could 'turn for stout-hearted help if disaster should descend',[72] wardens were to clear the streets, direct the public to shelters or refuge accommodation, ensure vehicles were drawn to the kerb so as not to impede the movement of fire engines, and tie horses and vehicles up. And, most infamously, they were required to call the 'attention of the occupier to any unobscured light in a building'.[73] But if wardens were not to succumb to the panic and turmoil of the situation that faced them, the correct personality was needed to be able to do 'the right thing in any emergency', and required, as keynotes of their conduct, 'courage and presence of mind' (Air Raid Precautions 1938a: 11).

In the current climate of terrorist and disaster preparedness, the civilian is again charged with what Jones et al. have called an 'indefinite vigilance' to await official instructions, but with far less responsibility for managing the outcome of the events (Jones et al. 2006: 59; see also 2004).

Entrainment

Burrowing down and below the surface has been a common response to an air threat in wartime art and architecture (Cocroft et al. 2003; Crowley and Pavitt 2008). During the Second World War, infrastructures of protection were both public and private: brick, stone, corrugated iron and other materials would protect people from the shell blast, fire and other effects. Families were encouraged to take responsibility by building their own shelters; even a trench could be dug in the back garden. ARP booklets (Association of Building 1938; Thomas 1939) prescribed the construction of a refuge room, an illustrated diagram advocating an intricate and amazing environment of withdrawal – a comforting spatiality that could wrap around the cowering populations with the basic necessities to survive.

Mattresses, paper pulp, air-tight jars, gas masks and overcoats allowed the occupants to take refuge from the turmoil outside. Food, books and toys, the wireless, would all permit life to go on as best as it could during the air raid, whilst mackintoshes, gum boots and galoshes, along with sand and water, a pick-axe and a shovel, were intended for the raid's aftermath – to deal with fires, with unexploded incendiary bombs, with collapsed structures or lingering gas. These reactive defensible spaces are evident today. For while critical infrastructures like airports act as targets for terrorism, we see efforts to 'design out' terrorism and 'build in' resilience with similar shelter or bunker architectures (Bosher et al. 2007; Coaffee et al. 2008).

Alongside the enveloping spaces of shelter, the aerial subject was to be buffered and insulated from the air raid by other technologies that were held closer to the body. A *prosthesis* of shelter developed. In one popular film clip an infant is enclosed and enwrapped by an all-body gas mask (Figure 7.4). The scene is watched by onlookers, whose scepticism is seemingly assuaged by the ARP officer explaining the mask's purpose and how it should be fitted. Like the pilots before them, the mask's temporary atmosphere protects

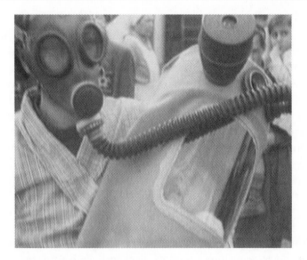

Figure 7.4   The interdependent gas mask 'chamber'.

the body from the environment made dangerous around them. But the mask further connects the baby's supply of oxygen to that of their mothers, maintaining the baby's life-grip upon the air and their parent.

As well as the material and maternal fabric of built structures and prosthetics, the bombed subject was advised to take shelter under an architectural array of instructions or preparations so that a calm and composed body could take shape. In Germany, 'psychological vulnerabilities' were calmed by 'political therapies' (Fritzsche 1992: 178). Drawing upon the values of experiential learning, staging and performative habituations and routines, aggression from the air was best reacted to by action without thought – or, rather more precisely, action entrained and habituated by a host of procedural and environmental factors which encouraged deliberation and reflection to occur before and away from the subject under siege. Entraining people to deal with a raid's 'atmosphere of terror and destruction' was vital.[74]

During the Blitz, the government's Home Intelligence Reports commissioned by the Ministry of Information found evidence of what had been described as 'moral maintenance'.[75] They picked up an anonymous 'working class district' where a large fire had endangered the community but a group of civilian fire watchers had kept an eye on it and done their utmost to keep the neighbourhood safe. For the reporter, 'despite the terrifying experience to which the residents had been subject during the raid, the following morning we found that the topic of conversation was not "the terror" but "the job" they had done'.[76] Being able to do something, the Home Intelligence reporters believed, 'was perhaps responsible for keeping their mind off the more terrifying aspects of the bombing'.[77] The air war needed to be addressed by much more than imagination or thought. *The Times* argued that 'the mere fact of having instructions to follow, duties to carry out, preparations to make, is the best antidote to fear'.[78] But these instructions had a dual role, of course. Incendiary bombs, gas attack and particularly mustard gas could be diffused without harm if dealt with quickly and orderly. The ARP devised many instruction leaflets and booklets that suggested how people could remove residue of the poison gas lewisite from their hands before burning, how they should wear their gas masks, how they could put out an incendiary bomb with speed. Without such preparations it was feared that 'an uninstructed, undefended population will certainly be thrown into panic'.[79] Defending an air raid was met by a series of preparations – processes of staging and rehearsing that took advantage of the group-herd mind disposed to panic yet also susceptible to fellow feeling.

The Air Raid Precautions booklets were written on the lines that households should prepare for a raid by turning their mind to developing a plan that would be put into motion on the event of a raid. The front cover of one such booklet read 'read this book through then keep it carefully' (Air Raid Precautions 1938b). To *keep* was the active verb: keeping carefully meant the continual maintenance of the book's safety and handiness. The manual was conceived in Parliament to be similar to the Highway Code in its capacities as a machinery of regulation and self disciplination.[80] Nothing was to be left to chance, nothing was to be done without prior contemplation. Readers were instructed to read through the notes and 'Think them out' and decide how they might be applied to one's own home. This would allow the reader to have 'faced the problem before danger threatens'. They will then know 'what to do if there should ever be a war', and, having made their plans, 'be ready to carry them out quickly if there is need' (Air Raid Precautions 1938b: 11).

In the networked social structure of a household, everyone was to be given a role. The head of the house should take command. Everyone would know what to do and where to go with the effect of 'no indecision and no panic. Risk is reduced to a minimum. Appointed tasks are undertaken, appointed places "manned" without hesitation and without confusion' (Air Raid Precautions 1938b: 16). In Germany, Nazi literature and novels emphasized a similarly instinctive citizen responsible for the *Volkskorper* or people's body: 'Be Quiet; Be Quick; "You know what you have to do", a character from Ernst Oligers tells her neighbours' (Fritzsche 1992: 213). Knowing that these plans had been carried out and that the possible had been faced was good enough for the government. Having been 'planned – and perhaps rehearsed', the knowledge could have a 'beneficial psychological effect on everyone, thus tending to avert the most fruitful source of disaster, namely, panic!' (Foulkes 1939: 78). As Fritzsche goes on, '[A]ir-readiness was moral uplift; people would once again hold their heads up higher, since they would have to keep an eye on the sky" (1992: 214).

John Langdon Davies (1939) advocated an approach very similar to the Home Office, suggesting that it was, again, a matter of facing the oncoming raids by finding the right mental frame. For Davies, the paraphernalia of 'shelters, gas masks, air raid wardens, drills' was not nearly as important as 'properly preparing your mind, and not exaggerating the danger, and of knowing how to get oneself in the right mental condition so not to be upset by the danger when it comes' (1939: 44). One way to do this was to develop procedures of movement and action. Metaphors took shape in Davies' book as Jim compared the situation to a desert island once again: 'If we sat around saying, "I'm afraid I shall be struck by lightning; I'm afraid I shall be eaten by a wild animal, we should have a bad time; but if we keep busy, none of these things will happen, and our nerves won't get on edge either."' For a workmate who had experienced the bombing in Barcelona, he recalled finding that if 'I was doing nothing, it was very hard not to believe that the next bomb was going to come my way; but, if I was busy, things went much better' (Davies 1939: 44).

Logistical lives

Wartime governments transferred disciplinary machinery capable of affecting the 'mobile or immobile mass' by breaking it down into a 'calculated practice of individual and collective dispositions, movements of groups or isolated elements, changes of position' (Foucault 1977: 166). During the Blitz, civilians who had prepared for the attacks were readied to snap into action – to dutifully wear their gas mask or escape to shelter. Repetition and simulation were key strategies of entrainment and immunization to fear, particularly drilling practices, instructions and exercises. Prescribed air raid precautions booklets and manuals recommended motions that resembled a sort of Taylorist rationalization (Cresswell 2006a) of the body's movements – intended not for efficiency, but supposedly to 'steel' oneself in the face of danger. 'Backgrounds of expectation' were forged through none too subtle procedural and sequential movements which opened out new modes of thinking and feeling not unlike that of military drill (Thrift 2000a: 36). Folding one's clothes and placing them on the dresser in readiness; pillows and sheets prepared for a spell in the shelter – these were two of many orderly examples people were encouraged to follow. Many found that they did them naturally as a way to prepare or react to the conditions they were forced to live with.[81]

In the event of a gas attack, ARP instructions advised how to put on a gas mask. Like the drilling of recruits of the Air Training Corps, the suggested movements were minutely prescribed by descriptions to be followed and practised. Accompanied by visual instructions, the booklet on gas went:

i)   Hold the breath.
ii) With the left hand grasp the bottom of the haversack and swing it to the front of the body. With the right hand remove any headgear and place it between knees.
iii) Thrust the right hand into the mouth of the haversack, and take hold of the respirator round the outside of the string binding, which secures the container to the facepiece. Withdraw the respirator with a smart movement and hold it up in front of the face ready for putting on. (Air Raid Precautions 1935: 79)

Stage (iv) was the most complex, requiring some manual dexterity as well as a the comprehension of very complicated instructions:

> Insert the thumb of the left hand under the centre of the headharness (i.e., at the point where all the elastic bands meet), release the grip with the right hand and allow the respirator to hang by the head harness on the left thumb. Insert the right thumb alongside the left thumb and slide the thumbs wide apart so that the elastic between them is stretched and the respirator is suspended by the middle and bottom elastics on each side. (Air Raid Precautions 1935: 85)

After this was completed the fairly easy task of thrusting the 'chin forward into the chin of the facepiece' could be undertaken; drawing the 'afterpiece on the face' would then require 'passing the head hardness smartly over the head with the thumbs'. All that was then needed was to make several adjustments to make sure the respirator was both comfortable and safe – ensuring the edges were not folded inwards or any of the elastic bands were twisted. One could then '[b]reathe out and continue to breathe in a normal manner' (Air Raid Precautions 1935: 85). Militarized regimentation such as this continued into instructions for first aid parties attempting to treat injuries, shocked members of the public, or even moving incapacitated and 'insensible' people from buildings.[82] The rational technologies of drill, dressage and operational research enabled poise in the face of an eventual air raid.

The response to contemporary threats continues in this vein. With the US readiness campaign, produced by design agency Ruder Finn, 'Get a kit', 'Make a Plan' and 'Be informed' mark a simple series of linear instructions designed to prompt actions, and preparations for those actions. Like the ARP measures, the UK government civil contingencies programme advocates a pictorial schematic of preparation, movement and action. The 'Preparing for Emergencies' booklet which was sent to every postal address in the country in 2004 educated its readers on what they needed to 'know' before an event. Their 'go in, stay in, tune in' slogan, pictured with coloured images, prescribed simple programmatic actions to be conducted by 'simple' people in the advent of an emergency. Reminiscent of the faceless airport PAX, the protagonist of the US instructions or 'airtoons' – the term 'coined by internet humorists to describe airline safety-card characters' – is described by Veronique Vienne (2004) as 'a white man wearing khaki pants and a polo shirt. A listless, emotionless "pod" figure.' Instincts and difficult to instruct feelings should still be trusted, however: 'Be Ready' suggests that one should be 'aware' of one's 'surroundings'. If something is amiss, you should 'move or leave if you feel uncomfortable or if something doesn't feel right'.[83]

Consider the lines, lanes and performances undergone in the airport security checkpoint. Enrolled into a script of security processing and removing one's shoes, to enact the 'small gestures' of taking off one's belt, turning out one's pockets of keys and change is to play out the role in the staging of what Tracy Davis (2002) has called the 'what if' (see also Aaltola 2005). The passenger embarks upon more narratives where security is assured by *doing*. As Davis shows in the context of cold war emergency preparedness exercises, the purpose of performances in these contexts cannot be underestimated as a way to 'train and, if possible, habituate people to react intellectually rather than emotionally in crises' (T. Davis 2002: 38). Exercises such as these originated in air raid drills.

Worrying about their disturbing and primitive effects upon urban terror, Lewis Mumford found that the air raid drills, whether 'arranged or real', produced 'similar effects' (cited in Saint-Amour 2005: 231). For Saint-Amour, 'the real war and the rehearsal for war become psychotically indistinct, nearly interchangeable backdrops before which the highly automated ritual of anticipation, dread, and mass-traumatization is enacted' (2005: 231). In the manner of earlier airshows, Air Raid Precautions exercises walked a careful line between spectacular performance and operational utility.[84] For the purposes of public display, ARP exercises needed to be 'well and realistically done, so as to avoid adverse comments either by the press or the public'.[85] In other local authorities, it was decided that the public training of recruits should not take place until a degree of 'efficiency has been reached that will ensure reasonable smooth working and competent standards of performance'.[86] Careful training was required before recruits could appear in public, for nothing could 'be so detrimental to the interests of air raid precautions than displays which are open to criticism or ridicule'. Reassurance was one outcome ARP exercises were intent on creating. Air raid drills at schools saw composed children evacuated to underground shelters, or calmly putting on their gas masks before quickly returning to their studies as if nothing had happened.

The conduct of public displays needed the right tone and pitch. The sense and atmosphere of the drill need to *be* spectacular and retain the audience's participation and captivation (Dent 2002). It was important that 'the performance should be organised, produced and timed so as to avoid awkward pauses or gaps when nothing is happening. The interest and attention of the audience must be held.' Whilst every display should be to some extent spectacular, considerations of realism appeared of the '*utmost* importance'.[87] Even if the displays were inevitably spectacular, what the drill or exercise was

Figure 7.5   How to combat a household fire. (Reproduced with permission of Corbis.)

intended to practise could be incredibly mundane or common sense. Public demonstrations were broadcast in cinemas in order to exemplify how to put out domestic fires (Figure 7.5), whether caused by an incendiary or not.

In one situation destined to be exhibited at a public display, the instructions for wardens focused on the timing of the incidents at which bombs would be expected to fall. If an attack was assumed to be made by a formation of aircraft, all its bombs would be dropped within two minutes. Why then space out the incidents to around 5–15 minutes, which would evacuate the situation of the quantitative and qualitative realism of bomb dropping? The situations were more than timings and patterns but the experiences these timings would create. 'Long delays between the incidents will not produce the pressure of work at the report centre which may be expected in war,' a memo suggested. 'There is every advantage in keeping exercises short and at the pressure likely to be experienced in war.' In other words, like emergency exercises today, the affect of the exercises had to be as realistic as the experience of war, so as to 'captivate' and 'motivate' (Schoch-Spana 2004).

Exactly where these exercises were to take place needed to be thought through with some care. Before a simulated air raid in Hackney, east London, the exercise committee thought seriously about the type of population they were experimenting on first, deciding not to carry out the experiment in Fulham for 'psychological reasons' that were not clear to this author. Caution was given to over-exciting the population or, indeed, scaring them into a panic. It would be a problem of crowd control, the public looking on at the display as if it was a 'a sort of free show', their congregation needing to be 'carefully controlled'.[88] Yet, whilst the reception of the exercises was anticipated, it was admitted that '[n]o one can say beforehand how the public will react to this experiment'.

Conclusion

A historical precursor to contemporary emergency preparedness exercises, the Blitz air raid drills were aimed at forming capabilities of readiness. Their concerns for realism and plausibility reflect many important considerations given to emergency planning exercises today. However, it was their capacity to produce dispositions that was their greatest weapon, to produce bodies 'disposed for action – "readied"'. As with the Air Training Corps, readiness or preparedness meant not the action itself, 'but a disposition toward a specific quality or form of action' (Crandall 2008).

As aerial life is attacked, so it must be defended. Unlike the security profile and the bomber target, which use imaginations of subjects in order to break down their enemies or individuate their objects into 'guilty', 'suspicious' or 'high-risk', the strategies for aerial protection we have seen reverse this operation. The reality is of sheltered societies protected not by tonnes of earth and concrete above one's head, but the engineering of collectives of people and things, whose task it is to tie their bonds ever closer together. A prosthesis of protection emerges in which people and everyday objects shelter one another from above.

In this social landscape, the terrain is short-circuited by communication, both verbal and affective. The family unit and civil society hold together the population's nerves. By movement and repetition, readiness is assured, entraining future movements so that they may be performed without thought.

Chapter Eight

Conclusion

The aerial body has been caught in the middle of so much: pummelled by high-intensity explosives and reverberated by voice broadcasts and sonic booms, it has been poked, pinched, bled and concussed; immersed and soothed, frisked and agitated. The aerial subject is overlain by legal rights and subjectified by various kinds of gaze and scrutiny. It has been made wealthy, courted as celebrity, bestowed with glamour and had its desires fulfilled. It has been stripped back and opened up, intensively searched, imagined and calculated, scanned, read and predicted. Trained and improved, uplifted and augmented, or even toughened and stretched, our aerial life has also been harmed and diminished. Fatigued and stressed, drowned, burned and killed, life has been all but snuffed out.

Stuck in the middle then, aerial life has been born through the regulations, technologies, practices, forces and affects that surround it and bear down upon on it. My argument has not been to detract from the symbolic life of the aeroplane, its passenger, voyeur, or pilot, but to show how the flesh and bones of the mobile body – at first so absent from the firework displays of 'Shock and Awe' in Iraq – that life of feelings, emotions, sensations and per-ceptions, has been fundamentally altered by the spaces and geometries of the aeroplane's movement. The aerial body is both the object and medium here. The conditions of its survival require that it is secured through all manner of techniques that, in turn, threaten the quality of that life.

Our understanding of the mobile body is given axis in this treatment, formed through imaginations of bodies conceived within the technique of the passenger profile, or targeted by the strategic bomber. With pasts and futures *projected* forwards and back, the mobile subject is understood by what it has done and what it might be about to do, how it might become dangerous, and how it is acted upon before that fact. Moreover, our aerial subject has been given depth, with a surface and an interior; intentions and desires may be read or incited by an address of superficial or far more vital capacities. It has not been the aim of the book to understand the body passively, however, as some inert surface to be stamped with an aerial signature. The aerial subject bounces back. It fights, flees, improves, toughens, and resists the poverty of the imaginations that attempt to govern it – from the fleshless PAX and the Malayan terrorist, to the airtoon characters. It gives way when it shouldn't and it exceeds the hand of power and the governance of intent.

The aerial body-subject is thus transformed and transforming of what I have argued are the shapes and spaces produced by the aeroplane – our aereality. From the space of the body to the biometric profile, from the geography of the blast explosion to the hologram of territory, and from the Amazon region to the monotopian vision of Europe, airspaces are uneven, distributed, vertical and horizontal. They are domains and doings, performing different shapes and geometries of insides and outsides. Furthermore, they are particularly environmental, vital and immersive.

Environments

The book has been chiefly concerned with the bodies of the aeromobile and it has significantly not had that much to do with its costs, at least in a traditional sense of the term. Some of the most important and critical questions concerning the aeroplane's future are whether it can be sustained (Bishop et al. 2003; Upham 2003), in terms of whether its environmental damage could be reduced, and whether we have enough resources to keep aircraft in the air. As Kingsley Dennis and John Urry (2009) suggest, the apparent flexibility and liquidity of mobile technologies such as the aeroplane are entirely dependent upon the liquidity of the resource oil – a commodity with a limited future. In this context, it would seem natural to discuss the effect of all of this movement upon the world's habitats and resources.

This book did something a little different. A point I developed was that aerial life has been about the production of *environments*. Clearly these environments involve the erosion of resources and the degradation of landscapes, but they are understood more in terms of points of articulation between the aeroplane and its subjects. As the book has turned towards the aeroplane's involvement in practices of war, terror and security, the relation between the environment and aerial life has been most acutely revealed. Being-in-the-air has meant that while the environment functions as the hinge-point of the subject's *silent* living conditions – their 'vital capacities' to live – it has become particularly vulnerable to analytical and material attack. War and terror, writes Sloterdijk, can only be understood in terms of their grasp of the environment 'from the perspective of its destruction' (2009: 28). From intelligence officers immersing themselves in the desert, the experiments of Nazi scientists at Dachau exposing their *Versuchpersonen* to extreme conditions, to the new calming security checkpoint and the sonic submersion of terror in Gaza, aerial environments are foregrounded backgrounds and targeted as life-supporting milieus.

It is perhaps because of both their pervasiveness to surround and their *persuasiveness* to affect the body in all manner of ways that aerial environments are so powerful, and even make 'continual existence impossible'. Capturing the thrill of flight in the visual spectacle and comforting surroundings of the airshow, the environment becomes a means for the *political*. Utopic messages move, collectives form, the most vital capacities of the subject are acted upon. These new aerial environments are posed in relation to futures. As the airshows demonstrated the aeroplane's potential for transportation, peace and new kinds of dominance, the aeroplane 'breaks open the implicits' by foreshadowing the present with a hint of what aerial life could be like.

Futures

Air travel has always been about the future. Viewing it as a prophesied object that would transform society, the 'winged gospel' (Corn 1983) held a messianic belief in the aeroplane, bestowing its movements with the ability to create peace and transcendental uplift. Anticipating the worst to come, governments moved into states of urgency in order to protect against the oncoming shadow of the bomber. Passenger and psychological profiles use past patterns of behaviour in order to project forwards. Decisions may be made over predicted intentions that turn over into actual and violent actions.

The aeroplane has promised a lot. It has promised peace and destruction for some. It has promised glamour and social capital, whilst it has guaranteed social as well as geographic mobility. It has promised employment and economic growth as well as new and more efficient ways of moving. Even the act of looking up at the sky is culturally understood as a gesture of forward thinking, or 'blue skies': the possibilities of something new coming along on the horizon. The aeroplane has promised a lot, an awful lot, and so it has a lot to answer for.

Indeed, it is upon these futures that the techno-cultures of the aeroplane's security, or violence, may act. States of readiness preparations for the Blitz resonate with contemporary preparedness machinery which attempt to manage how an aerial or other event will unfold. Efforts to divine the character of pilots and the intent of passengers rely upon indications of the future in order to act upon the present. Or, to be more precise, the future is often presented in order to prevent it from taking place or to manage the way it will unfold. Aerial life has therefore often been about anticipation. In our wonders and hopes, fears and terrors, the aeroplane's futures matter and they matter now.

But we can go even further and ask how this aereality can indicate what our everyday lives will look like. As Mark Salter has summed up, spaces such as the global airport work as 'a gauge of the power of the state and the relations between the state and society. The airport serves as a miniature example of other social processes of management and control' and as an example of the 'democratic and social implications of new modes of control and facilitation' (2008: 23). As Bulent Diken and Carsten Laustsen also note, 'in control societies, sociality seems to follow the pattern of the airport' (2006: 450). Even the future of the city may be the way of the airport.

Aerial Turns

Finally, this area of research gestures towards new promising fields of investigation. Given the diverse research drawn upon in this book – particularly within the social sciences – I want to suggest that we are witnessing not necessarily some kind of turn or new paradigm emerging, but at the very least a substantial opening up of the study of the air to critical thought and study. I mean this in the sense not of an instrumental focus on the aerial as a problem of governance, an issue of management routines and structures to be solved, but of the social, cultural and political relations which airspace and air travel can intersect, make or break.

What is clear in much of this research is the inextricable relationality between life in the air and life on the ground. We see the airport terminal emerging as a space examined for its demonstration of consumerism, identity or security protocols. From the air, the city emerges as a target. From great heights, weapons bury down to critical infrastructures. Airspace serves as an extra skein of territoriality manifesting supra-national geopolitical relations and treaties such as Europe's monotopia; it may even threaten the integrity of the nation-state. Aviation surveillance extends outwardly beyond the limits of the airport and airline systems. Dataveillance pulls and pushes information, collecting data regarding everyday life patterns, routines and consumer choices in order to inform an index of risk.

At the heart of this book is the affectivity of aerial life. The aeroplane has become a modulator of emotion and feeling, creating 'targets' of people and their collective sentiments. It harnesses rage and anger over its locations or disturbances. It can excite extreme thrills and passions, whilst it causes pain and disgust, fear and resentment. Air travel can strip us of our dignity by revealing vulnerabilities we may want to hide. For many it is incredibly fatiguing and dehumanizing. To be sure, just the simple 'quality of life' issues at stake by the aeroplane's motion deserve far more attention than they have.

Nothing is the focus of more collective sentiment than a subject such as the weather. But the sky, too, appears to be making a comeback in academic inquiry. As a site of connection of the physical and natural sciences with the social, recent studies of the representation of the sky in art and literature add much more to our social understandings of the weather. In explorations of the relations between the state, science and the status of knowledge (Whitehead 2009), the weather, being an object of risk and uncertainty, is increasingly examined beyond the laboratory as a domain of concern and action, from derivatives to the governance of climate change.

The sky's the limit.

Notes

CHAPTER ONE: INTRODUCTION

1 The meeting brought together Britain's leaders in military and civil aviation: Hugh Trenchard, then Chief of Air Staff, was in the midst of founding RAF (Cadet) College Cranwell; Frederick Sykes had been a year in post of Minister for Civil Aviation; and Geoffrey Salmond took the leading role of air commander for the Middle East.
2 'There can be no disputing the fact that the introduction of the aeroplane has created a new and different set of geographical values of great historical significance' (Balchin 1947: 6).
3 S. Stein, translation of document 1617-PS. Letter from Himmler to General Field Marshal Milch concerning transfer of Dr Rascher to the Waffen-SS (Office of United States 1946).
4 Ibid.
5 T. E. Lawrence to B. H. Liddell Hart, 26 June 1930, DG 694-5 (http://telawrence. org).
6 Maher Arar, statement, CBC News Online, 4 November 2003 (http://www.cbc. ca/news/background/arar/arar_statement.html).
7 Marinetti's *Manifesto dell'Aeropoesia* made 22 recommendations for a succesful aeropoem (see Bohn 2006: 209).

CHAPTER TWO: BIRTH OF THE AERIAL BODY

1 'Half-yearly inspection', *RAF Cadet College Magazine* 3 (2), 1922, p. 59.
2 This is of course in mind of the various discourses surrounding performativity, following the work of Judith Butler (1999), for whom subjectivity is a repetitive accomplishment.
3 For Gagen, 'the transformation of individuals was reconceived as a process attainable via the muscular conditioning of consciousness' (2004: 442).
4 *Air Scouting: General Scheme*, 1941, The Scout Association Archive (hereafter SAA).

5   Ibid.
6   'The early history of Air Scouting' (http://www.scouting.milestones.btinternet.
    co.uk/airscouts.htm).
7   *The Air League Bulletin* 5 (6), 1927, British Library (hereafter BL).
8   *The Air League Bulletin* 3 (53), 1925, BL.
9   Ibid., pp. 32–3.
10 *The Air League Bulletin* 4 (56), 1926, BL.
11 Ibid., p. 11.
12 'Air League Young Pilots Fund', Press Release, 1935. Royal Air Force Museum
    (hereafter AFM).
13 Letter from Chamier to Lord Brabazon, 2 January 1935, AFM.
14 Letter from Chamier to the Air Ministry, 19 November 1937, AIR 2/2716. The
    National Archives (hereafter TNA).
15 Letter from Chamier to the Air Ministry, 21 December 1937, AIR 2/2716,
    TNA.
16 'Air Cadets in LCC Schools Education Committee's refusal', *The Times*, 4 May
    1939, p. 10.
17 Elliott Mills' pamphlet *The Decline and Fall of the British Empire* (1905) is most
    exemplary of this thought (Mills in Hynes 1968: 25).
18 Similarly, Sidney and Beatrice Webb's social imperialism proposed that nation-
    alistic citizens were needed to ensure the Empire's survival. Such ideas impacted
    politicians and public figures such as Baden-Powell, who encouraged all to rush
    out and obtain of a copy of Elliott Mills' work (see note 17).
19 'The Wrights – a personal impression', *Flight Magazine*, 8 May 1909, p. 262.
20 Captain W. E. Johns, 'Handsome is as handsome does' *Air Training Corps Gazette*
    I (II), 1942, p. 9, BL.
21 Ibid., p. 9.
22 Ibid., p. 9.
23 Ibid., p. 9.
24 Ibid., p. 9.
25 Ibid., p. 9.
26 Letter from the Air League to the Air Ministry, 27 June 1938, TNA.
27 'Airmans Course for Scouts', *The Headquarters Gazette*, June 1912, p. 132, SAA.
28 Ibid, p. 132; A special issue of the magazine *The Scout* the following month also
    gave many details of how to build darts and kites, and information about aero-
    planes and airships. *The Scout*, July 1912. SSA.
29 ATC pamphlet, 1941. ED 136/326, TNA.
30 Letter from the Air League to the Air Ministry, 11 April 1939, AIR 2/4027,
    TNA.
31 Letter from Mr J. W. H. Ratcliffe to the Air Ministry, 26 June 1939, AIR 2/4027,
    TNA.
32 'Baghdad', *RAF Cadet College Magazine* 4 (2), 1923, p. 74, BL.
33 Letter from Charles Longman to Robert Baden-Powell, 16 February 1928.
    TSA.
34 Letter from Piers Howell, Secretary of the Scouts, to Lord Hampton, 27
    December 1929. TSA.
35 Ibid.
36 Ibid.

37 Letter from Lord Hampton to Sefton Brancker, 18 December 1929. TSA.

38 *The Scouter*, 1934, p. 137. TSA.

39 Letter from the Air Ministry to the Gliding Association, 2 September 1949. T 225/1282, TNA.

40 Air Training Corps and Women's Junior Air Corps (Code B, 10): ATC and Civil flying clubs: air experience 1943–1950, AIR 2/5699, TNA.

41 *Wing Tips: The Journal of the Midland Gliding Club* 1 (1), 1938. Birmingham Public Archives (hereafter BPA).

42 'The romance of the air', *The Air Training Corps Gazette* 1 (3), 1941, p. 33, BL.

43 AIR 2/100, TNA p. 4.

44 'The romance of the air', pp. 33–4 (see note 42).

45 *Air patrols: how to form and run a patrol in your troop*, 1936, SAA.

46 There are resonances here with the 'geographer citizens' whom Matless (1998) discusses.

47 N. MacMillan, 'Geography matters', *The Air Training Corps Gazette* 1 (12), 1941, pp. 6–7, BL.

48 Ibid., p. 7.

49 'On Cranwell', *The Air League Bulletin* 4 (67), 1926, p. 15, BL.

50 Ibid., p. 15.

51 L. Taylor, *Ground gen for airmen: a brief guide for airmen of all trades and grades to the interior economy of life in the Royal Air Force*, [London], 1941, BL.

52 'The RAF as a university', *The Air Training Corps Gazette* 1 (6), 1941, p. 30, BL.

53 Ibid., p. 30.

54 Ibid., p. 30.

55 Ibid., p. 30.

56 'The spice of life', *The Air Training Corps Gazette* 2 (4), 1942, p. 31, BL.

57 Ibid., p. 31.

58 Ibid., p. 31.

59 'An impression', *RAF Cadet College Magazine* 1 (1), 1920, pp. 20–1, BL.

60 Ibid., p. 20.

61 Board of Education, 'Recreation and physical fitness for youths and men', [London], 1937, BPA.

62 'The challenge of youth', Board of Education Circular 1516, 27 June 1940, BPA.

63 Ibid.

64 'Youth Service Corps', Board of Education Circular 1543, 12 March 1941, BPA.

65 Ibid.

66 Brunton's ideas can be found in the preface to Cantlie (1906). Brunton writes: 'a healthy brain must be lodged in a healthy body' (Cantlie 1906: xiii).

67 'Hints on athletic training', *RAF Cadet College Magazine* 6 (1), 1925, p. 34, BL.

68 'The new physiology', *The Times*, Wednesday, 30 May 1928, p. 15.

69 *Air Defence Cadet Corps Training Manual*, p. 21, BL.

70 Ibid., p. 21.

71 Ibid., p. 21.

72 Ibid., p. 21.

73   C. Wakefield, 'Sport as training for war', *The Air Training Corps Gazette* 2 (10), 1942, p. 2, BL.

74   C. Chamier, 'Commandant inspection: an impression', *The Air Training Corps Gazette* 2 (2), 1942, p. 29, BL.

75   Ibid., p. 29.

76   'Hints on athletic', p. 37 (see note 67).

77   *Manual of Training*, 2nd edition, September 1943. Issued from Air Ministry, Director of Air Training Corps.

78   Air League of the British Empire, *A drill book for the Air Training Corps*, 1943, BL.

79   'Hyde Park Rally of Youth', *The Times*, Monday, 27 September 1942, p. 6.

80   'Youth on Parade', *The Times*, Tuesday, 10 November 1942, p. 2.

81   Air League of the British Empire, *A drill book* (see note 78).

82   Ibid.

83   Ibid.

84   Ibid.

85   Air League of the British Empire, 1938, *Rules and regulations of the Air Defence Cadet Corps*, [London], 1938, p. 42, BL.

86   *The Scout's Pathfinder Annual* (The Scout Association, London, 1966), SAA.

87   'Cranwell: another point of view', *RAF Cadet College Magazine* 1 (1), 1920, p. 28, BL.

88   'Rifle shooting at Bisley', *RAF Cadet College Magazine* 4 (2), 1924, p. 82, BL.

89   'The joys of hunting', *RAF Cadet College Magazine* 2 (2), 1921, p. 85, BL.

90   'Hawking', *RAF Cadet College Magazine* 2 (2), 1921, p. 92, BL.

91   Beagling was eventually questioned by Hugh Trenchard, anxious for cadets not to be sidetracked.

92   'Hawking', p. 92 (see note 90).

93   'The young airman', *The Times*, Thursday, 26 September 1918, p. 3; *The Times*, Friday, 27 September 1918, p. 4; *The Times*, Tuesday, 1 October 1918, p. 5.

94   Wakefield, 'Sport as training for war', p. 2 (see note 73).

95   'Formation swimming seen as good training', *The Air Training Corps Gazette*, 1 (4), 1941, p. 36, BL.

96   *Air Scouting*, pamphlet 1944, SAA.

97   Ibid., p. 12.

98   Ibid., p. 16.

99   Ibid., p. 12.

100   *Aircrew exercises for the Air Training Corps*, Air Training Corps, London, 1943, p. 8, BL.

101   Ibid., p. 7.

102   'Practice theory and practice', *The Air Training Corps Gazette* 1 (5), 1941, p. 5, BL.

103   'Model flying', *The Air League Bulletin* 56, 1926, p. 16, BL.

104   Ibid., p. 16.

105   'Editorial', *The Skybird* 1 (1), 1933, p. 1, BL.

106   'Model flying', p. 17 (see note 103).

107   *Air scouting: organisation and training*, 1942, p. 13, SAA.

108   Ibid., p. 13.

109   'Model aeroplanes', M. R. Knight, *Empire Air Day, 1935*, brochure, p. 88, BL.

110 'Editorial', *The Skybird* 1 (1), 1933, p. 2, BL.
111 *The Skybird* 1 (2), 1933. pp. 37–8, BL.
112 'Letters', *The Skybird* 1 (2), 1934, pp. 37–8, BL.
113 'Photography of models', *The Skybird* 2 (7), 1935, p. 27, BL.
114 Ibid.

CHAPTER THREE: THE PROJECTION AND PERFORMANCE OF AIRSPACE

1   'Night scare, at the Cardiff docks, a weird whirr, startles coal-trimmers, airship with lights!' *The South Wales Echo*, 19 May 2009, Cardiff Public Archives (hereafter CPA).
2   'Editorial: The man about town', *The South Wales Echo*, 17 May 1909, CPA.
3   Jung draws upon the Nuremburg broadsheet of 1561: 'The obvious phallic comparison, i.e. the translation into sexual language, springs naturally to the lips of the people' (1979: 17).
4   For Paul Virilio (2005), the aeroplane and other forms of militarized transport are essential technologies of projection, ejecting projectiles that threaten the territorial body.
5   'Editorial: The man about town', *The South Wales Echo*, 22 May 1909, CPA.
6   Hendon Air Display Programme, 31 May 1913, p. 15, AFM.
7   Ibid., p. 15.
8   Ibid., p. 19.
9   Ibid., p. 19.
10 'Hendon and propaganda', *Flight Magazine*, 22 July 1920, p. 792, AFM.
11 Letter to Hugh Trenchard by Ethel Snowden, 13 July 1929, AFM.
12 Hendon Air Display Programme, 31 May 1913, p. 19, AFM.
13 Claude Graham-White, 'The public and aviation', reprinted from *The Country Gentleman and Land and Water*, 26 October 1912, p. 19, AFM.
14 Hendon Air Display Programme, 18 April 1914, p. 294, AFM.
15 Hendon Air Pageant Programme, 30 June 1923, TNA.
16 Letter to Hugh Dowding from Philip Game, 8 March 1922, AIR 2/4428, TNA.
17 Ibid.
18 'Training in the air', News cutting from *The Times*, 29 June 1926, AIR 2/4443, TNA.
19 'Agenda for conference on Empire Air Day arrangements', 25 June 1935, AIR 2, TNA.
20 Air Force Pageant Programme, March 1921, AIR 2, TNA.
21 Letter from W. S. Douglas to Hendon Organizers, 13 November 1926, AIR 2/4434, TNA.
22 Hendon Air Display Programme, 26 September 1912, p. 17, AFM.
23 'Air Pageant at the opening of Wolverhampton Airport', *The Express and Star*, 27 June 1938. Woverhampton Public Archives (hereafter WPA).
24 Letter from Air Commodore Chamier to Air Chief Marshal Sir Edward L. Ellington, Air Ministry, 25 October 1933, AFM.
25 Letter from W. L. Welsh to Air Officer Commanding, Inland Area, RAF Bentley Priory, Middlesex, 5 March 1934, AIR 2, TNA.

26   Letter from Air Vice-Marshal, Air Officer Commanding, RAF Halton, 22 March 1934, AIR 2, TNA.
27   'Empire Air Day: the question of insurance', 4 December 1934, AIR 2, TNA.
28   Letter from J. M. Spaight to Captain A. G. Lamplugh, 1934, AIR 2, TNA.
29   'Empire Air Day, advert, May 24th', published by the Air League, AIR 2, TNA.
30   'Empire Air Day 1937', Memo no. 3 1937, AIR 2, TNA.
31   'On arrangements for opening RAF Stations at home to the public on "Empire Air Day" Saturday, 25th May, 1935', April 1935, AIR 2, TNA.
32   Letter from Air Vice Marshal Commanding No. 22 AC Group to the Secretary Air Ministry DSD, 2 July 1937, AFM.
33   'Empire Air Day, 1934'. Report on opening of Royal Air Force Aerodromes and Stations to the Public, 29 October 1934, AIR 2, TNA.
34   'Empire Air Day' Souvenir Programme, produced by the *Air Review*, 1935, London, AIR 2, TNA.
35   Hendon Air Display Programme, 19 April 1913, p. 2, AFM.
36   Hendon Air Display Programme, 12 June 1913, p. 2, AFM.
37   Hendon Air Display Programme, 9 May 1914, p. 374, reprinted extract from *The Bystander*, AFM.
38   Ibid., p. 19.
39   Hendon Air Display Programme, 29 May 1913, p. 5, AFM.
40   Ibid., p. 5.
41   Hendon Air Display Programme, 13 May 1913, p. 19, AFM. A stretch of river known as the 'Ladies' Mile of the Thames', Boulter's Lock was an important centre for London high society. Upper class boating parties would gather on the river, whilst spectators viewed on from the bank.
42   Hendon Air Display Programme, 26 September 1921, p. 15, AFM.
43   Ibid., p. 15.
44   Hendon Air Display Programme, 9 May 1914, articles reprinted from the *Play Pictorial*, p. 374, AFM.
45   Schivelbusch's (1986) study of the railway journey was one of the most comparable examinations of the passenger, motion and spectator assemblage (see also Kirby 1997).
46   *Court Journal* excerpt in Hendon Air Display Programme, 9 May 1914, p. 374, AFM.
47   Ibid., p. 374.
48   Ibid., p. 374.
49   Cited in 'The 1910 Aviation Meeting at Wolverhampton', *The Midland Aeroclub Magazine*, 1910, WPA.
50   Ibid.
51   'A rebuff by the editor to the town's complaints', *The Express and Star*, 2 (58), 28 June 1910, p. 10, WPA.
52   Hendon Air Display Programme, 'A review of the 1913 season' by Charles Grey, 13 May 1913, p. 2, AFM.
53   Draft programme, 7 March 1921, AIR 2, TNA.
54   Hendon Air Display Programme, 26 April 1913, AFM.
55   Ibid., p. 9.
56   Ibid., p. 15.

57   'On the wings of the wind', extract from Miss Jaqueline Roberts' account of her flight at Hendon, Hendon Air Display Programme, 26 April 1913, p. 9, AFM.

58   Hendon Air Display Programme, 9 April 1913, p. 15, extract from *Tatler*, AFM. This description constrasts with John Urry's (2000) analysis of the car as some kind of bubble, distancing its driver from the outside. On the other hand, Merriman (2007) and Thrift (2004a) draw analysis that describes the car's ability to direct the driver's perceptions towards the road and other drivers.

59   'The Air Conference on the Necessity for Municipal Airports', *Flight Magazine*, 22 November 1929, p. 1191.

60   Ibid., p. 1192.

61   Ibid., p. 1197.

62   Ibid., p. 1198.

63   Ibid., p. 1198.

64   Appendix A. Finance and General Purposes Committee, Liverpool Council. H352 COU Liverpool Records Office.

65   'The Air Conference', p. 1195 (see note 59).

66   'The new Birmingham Airport', *Flight Magazine*, 6 July 1939, pp. 55–6.

67   Birmingham Airport Booklet, 1939, p. 19, BPA.

68   R. Wallace, 'The right to fly', *Flight Magazine*, 22 November 1913, p. 1280.

69   Interim Report of the British Delegates to the International Conference on Aerial Navigation, CAB 18/26, TNA.

70   Next Gen – The sister initiative from across the Atlantic – For Global Harmonization, European Organization for the Safety of Air Navigation EUROCONTROL, October 2008 (http://www.eurocontrol.int/sesar/gallery/content/public/docs/sesar%20in%20brief_2008_final%20low.pdf).

71   A Step Forward On Single European Sky, IATA Press Release, 25 June 2008 (http://www.iata.org/pressroom/pr/2008-06-25-03.htm).

72   Single European Sky Interoperability Regulation, Single European Sky Workshop, France, 17–18 December 2008; A Comparative Assessment of the NextGen and SESAR Operational Concepts, Joint Planning and Development Office (JPDO), Global Harmonization Working Group, 22 May 2008, Paper no 08-001.

73   Treaty on the Functioning of the European Union, *Official Journal of the European Union*, 12 December 2007 (http://eur-lex.europa.eu/LexUriServ/site/en/oj/2007/c_306/c_30620071217en00420133.pdf).

74   House of Lords, 'The Passenger Name Record (PNR) Framework Decision', 15th Report of Session 2007–2008, European Union Committee. London: TSO, 2008.

75   Regina v. Immigration Officer at Prague Airport and Another, House of Lords, Session 2004–5, on appeal from: [2003] EWCA Civ 666 2004 WL 2790691, before: Lord Bingham of Cornhill, Lord Steyn, Lord Hope of Craighead, Baroness Hale of Richmond, Lord Carswell, Thursday, 9 December 2004.

76   See R (ex parte European Roma Rights Center et al.) v. Immigration Officer at Prague Airport and another (UNHCR intervening) (2005), 17 IJRL 427 at 436-40 (Amicus brief filed by UNHCR).

77   '1963–2003: 40 years of service to European aviation', EUROCONTROL (http://www.eurocontrol.int/corporate/gallery/content/public/docs/pdf/aboutus/history.pdf).

78   Gaza–Jericho Agreement, Annex I, 4 May 1994.
79   'Netanyahu speech – key excerpts', BBC Online, 14 June 2009 (http://news. bbc.co.uk/1/hi/8099749.stm).
80   See the CENSIPAM/SIVAM website portal, Amazonian Protection System: SIPAM, (http://www.sipam.gov.br/en/index.php).

CHAPTER FOUR: AERIAL VIEWS: BODIES, BORDERS AND BIOPOLITICS

1   'Aviation in Parliament', *Flight Magazine*, 6 November 1919, p. 1460.
2   Captain H. Hamshaw Thomas 'The future of aerial photography in the East', in 'Aeroplane photography for map making', 1919, p. 26, WO 181/446, TNA.
3   Ibid., p. 26.
4   'An investigation of the possibilities attaching to aerial co-operation with survey, map-making and exploring expeditions', Gordon Shephard Essay Prize Winner, 1922, AIR 1/161/15/123/19, TNA.
5   Colonial Office Conference, 19 May 1927, 1 pm, p. 17, CO 323 990 7, TNA.
6   Ibid., p. 17.
7   'Notes on surveys by aeroplane photography suggested by Brig. Gen. J. D. Hearson's evidence before the Colonial Survey Committee and the subsequent discussion', 1919, CO 323/807, TNA.
8   'An investigation' (see note 4).
9   Lt-Col. T. E. Lawrence, 'Map making by air photography', *Daily Telegraph*, 30 October 1920.
10   Colonial Office Conference (see note 5).
11   'An investigation' (see note 4).
12   Report by the Conservator of Forests on the Valuation of the Forests of the Bartica-Kaburi Area, British Guiana Combined Court, Second Special Session, 1926, CO 111/668/15, TNA.
13   Ibid.
14   Sir Thomas H. Holland, introduction from the chair, 'Air survey and Empire development', *Journal of the Royal Society of Arts*, December 1928, p. 162, TNA.
15   Colonel H. L. Crosthwaite, ibid., p. 162.
16   Colonial Office Conference, p. 8 (see note 5).
17   M. S. Mackey, 'Report on possibility of aerial survey and forest-valuation in British Guiana', Enclosure to British Guiana Despatch, no. 95, 14 March 1927, CO 111/668/15, TNA.
18   B. R. Wood, ibid.
19   Ibid.
20   Colonial Office Conference, p. 13 (see note 5).
21   Mr Oliphant in ibid., p. 9.
22   Letter from the Acting Conservator of Forests to the Colonial Secretary, 21 July 1927, CO 111/668/15, TNA.
23   'Report on an aerial forest reconnaissance in the North West District', report by M. S. Mackay, 18 August 1927; copy of minute by L. S. Hohenkerk, Esquire, Acting Conservator of Forests, 22 August 1927, my emphasis, CO 111/668/15, TNA.

24   'An investigation' (see note 4).

25   Holland, 'Air survey and Empire development', p. 169 (see note 14).

26   Captain F. S. Richards, 'Mapping from aeroplane photographs: with special reference to the methods of 7th Field Survey Company, R.E., on the Palestine Front', 1918, WO 181/446, TNA.

27   Captain H. Hamshaw Thomas, 'Aeroplane photography for map-making: experience gained by the Royal Air Force in Palestine and Sinai during 1917 and 1918', 1919, p. 14, WO 181/446, TNA.

28   Letter to Lord Passfield, Colonial Office, from the Governor's Office of Northern Rhodesia, 19 September 1929, p. 6, CO 795 32 12, TNA.

29   ICAO, Machine Readable Travel Documents Supplement to Doc. 9303, 2008 Version: Release 7, Status: Final Date – 19 November (http://www2.icao.int/en/mrtd/Pages/default.aspx).

30   'MRTDs: history, interoperability and implementation', ICAO, Version: Release 1, Status: Draft 1.4, Date: 23 March 2007, p. 22 (http://www2.icao.int/en/mrtd/Pages/default.aspx).

31   Ibid., p. 26.

32   'MRTDs: history, implementation and interoperability', p. 27 (see note 30).

33   Biometrics Deployment of Machine Readable Travel Documents, ICAO TAG MRTD/NTWG Technical Report Version 2.0 (http://www2.icao.int/en/mrtd/Pages/default.aspx).

34   Ibid.

35   Letter to Colonel Sir Charles Close, 27 January 1920, OS 1/11/1, TNA.

36   Ibid.

37   Letter from H. J. Winterbotham to Musgrave, 25 March 1920, OS 1/11/1, TNA.

38   Letter to Lennox Cunningham, 27 September 1919, OS 1/11/1, TNA.

39   Richards, 'Mapping from aeroplane photographs' (see note 26).

40   Thomas, 'Aeroplane photography for map-making', p. 14 (see note 27).

41   'Report on simultaneous photography by a flight of 3 aircraft on the Bajaur Frontier Area, February 1932', Indian Air Survey Committee, WO 181/34, TNA.

42   http://www.tsa.gov/evolution/.

43   4 September 2003, 7:43 a.m., in USVISIT Redacted Message to EPIC (http://www.dhs.gov/xlibrary/assets/usvisit/US_VISITRedacted_Message_for_EPIC.pdf).

44   Recommended Security Guidelines for Airport Planning, Design and Construction (2006), Transport Security Administration (http://www.tsa.gov/assets/pdf/airport_security_design_guidelines.pdf).

45   'Kano survey school by the acting Surveyor General and the acting Director of Education', 2 April 1913, OS 181/160, TNA.

46   W. R. H. Rowe, Surveyor General, 'Memorandum on the training of native surveyors in Nigeria Gold Coast and Sierra Leone', OS 181/160, TNA.

47   Lawrence, 'Map making by air photography' (see note 9).

48   Captain H. Hamshaw Thomas, 'Notes on aerial photography in relation to survey work with reference to future developments', 28 May 1919, p. 1, OS 181/160, TNA.

49   Ibid., p. 7.

50   Ibid., p. 1.
51   US-VISIT and You, What to Expect When Visiting the United States, Department of Homeland Security (http://www.dhs.gov/.../usvisit_edu_traveler_brochure_printer_friendly_english.pdf).
52   Letter from H. J. Winterbotham to P. E. Hodgson, 14 November 1924, WO 181/36, TNA; letter from Sgt P. Gethin to Colonel Hutchinson, 36 October 1928, WO 181/259, TNA; Captain H. Hamshaw Thomas, 'Notes on the use of photography in connection with transcontinental flying and air routes', 28 May 1919, TNA.
53   Colonel Fletcher, Commandant, Central Flying School, 'Stereoscopic photography from aeroplanes', 1916, WO 181/503, TNA.
54   R. Guy, 'Notes on stereo photography as applied to aerial (aeroplane) work', 1916, WO 181/503, TNA.
55   Ministry of Home Security, research and experiments department. 'Stereomarks' by T. A. Littlefield, WO 181/503, TNA.
56   Ibid., p. 8.
57   'Allied Central Interpretation Unit: plotting history', 1941, AIR 38/84, TNA.
58   Major H. Hemming, 'Air Surveying in Northern Rhodesia', *AIRWAYS*, February 1927, p. 2.
59   Letter to Secretary Janet Napolitano, DHS, 31 May 2009, EPIC Backscatter resources (http://epic.org/privacy/airtravel/backscatter/Napolitano_ltr-wbi-6-09.pdf).
60   Ibid.

CHAPTER FIVE: PROFILING MACHINES

1   House of Lords, Judgments – R (on the application of Gillan (FC) and another (FC)) (Appellants) v. Commissioner of Police for the Metropolis and another (Respondents), 8 March 2006 (http://www.publications.parliament.uk/pa/ld200506/ldjudgmt/jd060308/gillan-1.htm).
2   Ibid.
3   Ibid.
4   Ibid.
5   Ibid.
6   Boeing's Chief of the Equipment Unit in a report to the Flying Personnel Research Committee (hereafter FPRC), 'Altitude conditioning of aircraft cabins', FPRC 244, 1941, p. 1, AIR 57/3, TNA.
7   Dr Mackworth, 'The psychological effects of a loss of sleep', Psychological Research Committee 202, 1940, p. 1, AIR 57/3, TNA.
8   G. C. Drew, 'An experimental study of mental fatigue', FPRC, 1940, p. 14, AIR 57/3, TNA.
9   Ibid., p. 15.
10   Ibid., p. 14.
11   Ibid., p. 14.
12   Air Vice Marshall Whittingham, 'Visits of medical psychologists to stations', October 1939, p. 1, AIR 57, TNA.
13   Ibid., p. 1.

14 Ibid., p. 1.

15 Report of the Committee on the Preliminary Education of Candidates for Royal Air Force Commissions, 1921, p. 1, AIR 2/100, TNA.

16 Apendices III – Physical Tests of Fitness for Flying (evidence of Lt-Col. Martin Flack), p. 9.

17 Report of the Committee on the Preliminary Education of Candidates for Royal Air Force Commissions, AIR 2/100, p. 2, TNA.

18 Lt-Cdr F. Warren Morrison, 'Pre-selection of pilots', FPRC, 19 December 1940, p. 1, AIR 57/3, TNA.

19 Ibid., p. 2.

20 Ibid., pp. 2–3.

21 'A progress report on pre-selection of flying personnel', FPRC 71, 1939, AIR 57/1, TNA.

22 Ibid.

23 Ibid.

24 Prof. F. C. Bartlett, 'Temperamental characteristics and flying ability', FPRC 13d, 1939, p. 1, AIR 57/4, TNA. The cost of preliminary training added up to £400 per candidate. With 30 percent wastage and an entry of around 2,000–3,000 in 1943, around 600–900 men were receiving training not of 'direct use to the RAF' and at the cost of around £240,000–£360,000 per year. Prof. F. C. Bartlett, 'Psychology', FPRC 13e, 1939, p. 3, AIR 57/4, TNA.

25 Letter to Sir Frederick Banting from Professor E. A. Bott, FPRC 78c, 1940, p. 1, AIR 57, TNA.

26 Ibid., p. 1.

27 'The medical care of flying personnel in war', FPRC 265, 1941, p. 2, AIR 57/4, TNA.

28 Captain C. P. Symonds, 'Report of visits to stations in coastal command to investigate neuropsychiatric problems', 26 January 1940, p. 1, AIR 57, TNA.

29 'The medical care', p. 3 (see note 27).

30 Ibid., p. 9.

31 Group Capt. Symonds, 'Proposals for psychiatric assistance', Head Injuries Centre, St Hugh's College, Oxford, 3 April 1942, AIR 57, TNA.

32 Air Commodore H. L. Burton, 'Psychological selection of WAAF personnel', 6 May 1942, AIR 57, TNA.

33 K. A. Carmichael, 'Neurology', FPRC 13, 1939, p. 1, AIR 57, TNA; letter to Air Marshall Whittingham from Group Capt. Symonds, Oxford, 16th March 1942, AIR 57, TNA.

34 Group Capt. Symonds, 'Proposals for psychiatric assistance', Head Injuries Centre, St Hugh's College, Oxford, 22 March 1942, FPRC 13, AIR 57, TNA.

35 Burton, 'Psychological selection' (see note 32); 'The medical care', p. 8 (see note 27).

36 Group Captain R. D. Gillespie, 'Indications for referral of recruits to the psychiatrist at recruit depots', 8 September 1942, pp. 1–2, AIR 57, TNA.

37 'Notes of meeting of a Sub-committee to discuss airsickness among flying personnel', FPRC 220, 1941, AIR 57/3, TNA.

38 United States v. McCaleb (1977), United States of America, Plaintiff-Appellee, v. Robert Ross McCALEB and Brenda Page, Defendants-Appellants. Nos

76-2060, 76-2061. United States Court of Appeals, Sixth Circuit. Argued 15 December 1976. Decided 11 April 1977.

39   United States v. Van Lewis (1977), United States of America, Plaintiff-appellee, v. William Van Lewis, Defendant-appellant United States Court of Appeals, Sixth Circuit. – 556 F.2d 385, 6 June 1977. Rehearing denied 5 August 1977.

40   Review of the Operation of the Prevention of Terrorism (Temporary Provisions) Act 1989, Home Office 2000.

41   Report on the Operation in 2005 of the Terrorism Act 2000, Lord Carlisle of Berriew QC, May 2006.

42   'Committee Paper', FPRC, 12 Februrary 1941, p. 2, AIR 57, TNA.

43   A. W. Callaghan, 'High-altitude flying (its effects on the cardio-respiratory system and physical efficiency)', FPRC 29, 1941, AIR 57, TNA.

44   J. R. Poppen, US Navy Medical Corps under direction of C. K. Drinker of Harvard, 'Physiological effects of sudden changes in speed and direction in airplane flight', FPRC 263a, 1941, AIR 57/4, TNA.

45   Ibid., p. 1.

46   Ibid., p. 4.

47   Ibid., p. 5.

48   Letter Addressed to Kingsley Wood, from the Bioglan Laboratories Ltd, 16 February 1940, p. 1, AIR 57, TNA.

49   Dr D. R. Davis, 'Investigation into the psychological effects of Benzedrine on normal adults', FPRC, 1940, p. 3, AIR 57/2, TNA.

50   Drew, 'An experimental study', p. 45 (see note 8).

51   Ibid., p. 22.

52   Whittingham, 'Visits of medical psychologists', p. 1 (see note 12).

53   C. H. Hartley and Dr J. L. Young, 'Advantage of the use of red illumination in aircraft cockpits', FPRC, February 1941, p. 2, AIR 57/3, TNA.

54   M. D. Vernon, 'The ability to distinguish dimly illuminated aeroplane silhouettes', January 1941, FPRC 247, p. 4, AIR 57/4, TNA.

55   P. C. Livingstone, 'Ophthalmology: resumé of work done to date', FPRC 13, p. 2, AIR 57/4, TNA.

56   Ibid., p. 5.

57   Squadron Leader C. G. Williams, 'A method of studying coordination of binocular movement with special reference to flying', March 1939, FPRC 15, p. 2, AIR 57/4, TNA.

58   On stereoscopy, Saville-Sneath's instructions were reminiscent of Hotine's suggestions for photogrammetric techniques (Hotine 1927: 71).

59   K. J. W. Craik, 'Progress report on dark adaptation and night vision', FPRC 289, 1941, p. 8, AIR 57/4, TNA.

60   Ibid., p. 10.

61   Ibid., p. 7.

CHAPTER SIX: AERIAL ENVIRONMENTS

1   Hansard, 28 September 1934, vol. 295, col. 859.

2   Cited in R. Brooke-Popham, 'Some notes on morale', 1923, AIR 69/3, AFM.

3   Ibid., p. 1.

4 Ibid., p. 5.
5 R. Brooke-Popham, 'Exercise 14', RAF Staff College, 1924, AIR 69/36, AFM.
6 Le Bon in Brooke-Popham, 'Some notes' (see note 2).
7 R. Brooke-Popham, 'Remarks on Exercise no. 14', RAF Staff College, 1924, p. 4, AIR 89/36, AFM.
8 Ibid p. 2.
9 'Espirit de corps in the RAF' Conference number 5, RAF Cadet College, 1929, AIR 69/61, AFM; E. Edwards, 'Leadership and morale', 1936, AIR 69/129, AFM; M. Barrett, 'Leadership and morale', 1937, AIR 69/157, AFM; G. Beamish, 'Leadership and morale', 1937, AIR 69/146, AFM.
10 Tarde's ideas of imitation would later influence Émile Durkheim's (1997) studies of collective and contagious affect as a source of suicidal behaviours.
11 War Office Committee, Report of the War Office Committee of Enquiry into 'Shell Shock', London: HMSO, 1922.
12 'The variation of blast pressure in the vicinity of a cylindrical pellet of explosive II', Department of Scientific and Industrial Research, Road Research Laboratory, RC269, 1941, HO 195/12, TNA.
13 A. N. Black, B. Delisle Burns and S. Zuckerman, 'An experimental study of the wounding mechanism of high velocity missiles', Ministry of Home Security, RC 264, 1941, HO 195/12, TNA.
14 Ibid., p. 2.
15 Ibid., p. 2.
16 Ibid., p. 2.
17 Ibid., p. 1.
18 Ibid., p. 3.
19 'Model tests on the propagation of blast in straight streets', Report for the Ministry of Home Security, Department of Scientific and Industrial Research, Road Research Laboratory, RC 53, 1940, HO 195/7, TNA.
20 Ibid., p. 13.
21 J. D. Bernal, 'Principles of shelter policy', Home Security Air Raid Precautions Department, Research and Experiments Branch, RC 94, 1940, HO 195/7, TNA.
22 Ibid., p. 2.
23 J. D. Bernal and F. Garwood, 'An investigation of probable air raid damage', Ministry of Home Security, 1940, HO 195/7, TNA.
24 R. B. Fisher, P. L Krohn and S. Zuckerman, 'The relationship between body size and the lethal effects of blast', Ministry of Home Security, RC 284, 1941, p. 7, HO 195/12, TNA.
25 Letter from Harold Scott to Sir Henry French, 1941, HO 199/453, TNA.
26 *Evidence in Camera*, HMSO, 1945, p. 20.
27 'Investigation of morale and industrial effects of enemy bombing', Ministry of Home Security, Research and Experiments Department, HO 199/453, TNA.
28 S. Zuckerman, 'Observations on the effects of air-raids on Tripoli town and harbour, on shipping and on military targets in the field', Ministry of Home Security, 1943, HO 191/189, TNA.
29 Ibid.
30 Ibid.
31 Ibid.

32 Director of Operations, 'The destruction of CT cultivated areas', Instruction 27, 12 August 1953, AIR 23/8592, TNA.
33 Ibid.
34 'Directive of the Central Politbureau on Clearing and Planting', 25 September 1951, translated by the War Office, AIR 23/8592, TNA.
35 Ibid.
36 Ibid.
37 D. F. Bayly-Pike, 'The probabilities of success in area bombardment', Operational Research Unit, Far East, MEMO no. 2/57, 1957, p. 1, AIR 291/1701, TNA.
38 Ibid., p. 5.
39 Ibid., p. 6.
40 Ibid., p. 1.
41 Ibid., p. 1.
42 'Observations on the human ecology of the Malayan jungle', Operations Research Section (Malaya) Memo, 1954, WO 291/1742.
43 Ibid.
44 Ibid.
45 'Colonial affairs', House of Commons Debates, 17 July 1952, vol. 503, cols 2345–461.
46 'Jungle defoliation', House of Commons Debates, 23 April 1952, vol. 499, col. 396.
47 'Observations on the human ecology' (see note 42).
48 Ibid.
49 Ibid.
50 Ibid.
51 'Communist terrorist camps in Malaya', Operational Research Section (Malaya), Memo 3/53, 1953, p. 3, WO 291/1728, TNA.
52 Ibid., p. 3.
53 'Observation on the tracking abilities of Malayan aborigines', Operational Unit Far East, Memo 4/54, 1955, p. 13. WO 291/1689, TNA; 'Spotlight on the Emergency', notes from 'Sunday Morning Papers', 16 March 1952, CO 1022/57, TNA.
54 'Observation on the tracking abilities', p. 5 (see note 53).
55 Dr Eyad El-Sarraj, Psychiatrist, Director General Gaza Community Mental Health Programme Medical Report, 2006.

CHAPTER SEVEN: SUBJECTS UNDER SIEGE

1 'Cars and sirens: confusion between cars changing gears and air raids', 27 August 1940, p. 3, MO 371, Birmingham Public Library (hereafter BPL).
2 'Citizen guidance on the Homeland Security Advisory System' Department of Homeland Security (http://www.dhs.gov/xlibrary/assets/citizen-guidance-hsas2.pdf).
3 Homeland Security Presidential Directive 3: Homeland Security Advisory System (http://www.dhs.gov/xabout/laws/gc_1214508631313.shtm).

4 'Public warnings and air raid shelters', 14 October 1926, Subcommittee on Air Raid Precautions, Committee of Imperial Defence, Memo Papers 1926–8, p. 3, CAB 46, TNA.

5 Ibid., p. 3.

6 Ibid., p. 4.

7 Report of a Committee on 'Loud Noise' Air Raid Warning Signals, Home Office ARP Department, 1938, HO 451/897, TNA.

8 Letter to G. D. Kirwan, Home Office ARP Department, from Dr Tucker, Air Defence Experimental Establishment, Biggin Hill, Kent, HO 451, TNA.

9 Ibid., p. 5.

10 Report of a Committee (see note 7).

11 Ibid.

12 Letter to County and Town Clerks, 'Air raid warning signals', April 1938, from the ARP Department, Home Office, HO 451/897, TNA; purchasing advice given by ARP Division to use sirens made by Carter & Co. (Nelson) Ltd, William Streets, Lancs; ARP Department Circular no. 24, 'Types of instruments for air raid warnings', 6 February 1939, HO 451/897, TNA.

13 Letter to County and Town Clerks (see note 12).

14 'Public warnings', p. 4 (see note 4).

15 Ibid., p. 4.

16 Ibid., p. 4.

17 Ibid., p. 6.

18 Ibid., p. 6.

19 Ibid., p. 6.

20 Ibid., p. 9.

21 Ibid., p. 10.

22 Ibid., p. 18.

23 Mass Observation Report, 23 August 1940, p. 1, MO 364, BPL.

24 Ibid., p. 2.

25 'Public warnings', p. 21 (see note 4).

26 Public Notice, Air Raid Warning Test, Saturday, 16 March 1940. Birmingham City Archives.

27 Air Subcommittee of the Cabinet Office, 16 February 1931, CAB 46, TNA.

28 Air Subcommittee of the Cabinet Office, 20 May 1931, p. 14, CAB 46, TNA.

29 Samuel Hoare, Hansard, House of Commons Debates, 15 November 1937, vol. 329, col. 43.

30 Ibid., col. 41.

31 Air Subcommittee of the Cabinet Office, 13 February 1931, p. 14, CAB 46, TNA.

32 Air Subcommittee of the Cabinet Office, 20 May 1931, p. 14.

33 House of Commons Debates, 15 November 1937, vol. 329, col. 137.

34 Minutes of the Meeting of the Subcommittee, 9 February 1925, Subcommittee on Air Raid Precautions, Committee of Imperial Defence, p. 1, CAB 46/1, TNA.

35 Ibid., p. 4.

36 Ibid., p. 3.

37 Ibid., p. 13.

38   Air Subcommittee, 16 February 1931, p. 16.
39   Air Subcommittee, 20 May 1931, p. 15.
40   Air Subcommitee, 16 February 1931, p. 18.
41   Ibid, p. 18.
42   Ibid, p. 15.
43   Air Subcommittee of the Cabinet Office, 18–23 March 1931, p. 15, CAB 46, TNA.
44   Ibid., p. 15.
45   Ibid., p. 16.
46   Final minutes of meeting held at the ARP Department, 18 October 1938, Committee on Smoke Screens, HO 45/17595, TNA.
47   'Notes on possible use of smoke for camouflage', 1 June 1938, for consideration by the Committee on Smoke Screens, ARP Department, HO 45/17595, TNA.
48   P. R. C. Groves cited in Professor P. J. Noel Baker, 'Foreign affairs and how they affect us', broadcast 22 February 1927, transcript produced in Subcommittee on Air Raid Precautions Memo Papers 1926–8, p. 2, CAB 46, TNA.
49   Ibid., p. 2.
50   House of Commons Debates, 15 November 1937, vol. 329, col. 55.
51   Ibid., col. 114.
52   Lt-General Sir Gerald Ellison in a letter to *The Times*, 29 December 1938, p. 53.
53   House of Commons Debates, 15 November 1937, vol. 329, col. 151.
54   Ibid., col. 89.
55   http://www.Ready.gov.
56   Ibid.
57   House of Commons Debates, 15 November 1937, vol. 329, col. 44.
58   Ibid., col. 44.
59   ARP Department, 'The working of a control centre', 30 August 1939, BPL.
60   Ibid., p. 2.
61   Ibid., p. 3.
62   'Duties and training of air raid wardens', a circular sent by city of Birmingham, extracts taken from memo no. 4, issued by the Home Office (ARP Department), HMSO, BPL.
63   Ibid.
64   Ibid.
65   *Birmingham Mail*, 10 January 1940, BPL.
66   'Rumours about air raid damage and causalities', Civil Defence Executive and Subcommittee, 4 June 1942, HO 186/886, TNA; memo for the Prime Minister, 17 February 1941, Minister of Home Security, HO 186/886, TNA.
67   Oral history tape, Ellen Ada Harris (1987), The Imperial War Museum Sound Archive 9820, Imperial War Museum, London.
68   Ibid.
69   Ibid.
70   Ibid.
71   'Co-operation between Mental Health Emergency Committee and civil defence services', 1941, HO 287/307, TNA.
72   'Duties and training of air raid wardens', p. 3 (see note 62).

73 Ibid., p. 11.
74 Parliamentary Air Raid Precautions Committee Memo, 28 August 1941, pp. 2–3, HO 186/970, TNA.
75 Home Intelligence Report, 14 March 1941, BPL.
76 Ibid.
77 Ibid.
78 'ARP things that need knowing, III. – contrasts in a vast experiment', *The Times*, Wednesday, 12 January 1938, p. 11.
79 Ibid.
80 House of Commons Debates, 18 June 1936, vol. 313, col. 1154; see also Merriman (2007) on the governmental technology of the Highway Code.
81 Imperial War Museum Sound Archive 14595, Imperial War Museum, London.
82 Air Raid Precautions Memorandum, No. 10, 'The training and work of first aid parties', 1939, BPL.
83 http://www.Ready.gov.
84 Air Raid Precautions Memorandum, No. 9, Notes on 'Training and exercises', 1939, p. 1, BPL.
85 Ibid., p. 10.
86 Ibid., p. 2.
87 Ibid., p. 10.
88 Parliamentary Air Raid Precautions Subcommittee, 20 February 1933, p. 6, CAB 46/29, TNA.

Bibliography

Aaltola, M. (2005) 'The international airport: the hub-and-spoke pedagogy of the American empire', *Global Networks* 5 (3): 261–78.

Aas, K. F. (2006) '"The body does not lie": identity, risk and trust in technoculture', *Crime, Media and Culture* 2 (2): 143–58.

Adey, P. (2004a) 'Surveillance at the airport: surveilling mobility/mobilising surveillance', *Environment and Planning A* 36 (8): 1365–80.

Adey, P. (2004b) 'Secured and sorted mobilities: examples from the airport', *Surveillance and Society* 1 (4): 500–19.

Adey, P. (2006) 'Airports and air-mindedness: spacing, timing and using Liverpool Airport 1929–39', *Social and Cultural Geography* 7 (3): 343–63.

Adey, P. (2008a) 'Architectural geographies of the airport balcony: mobility, sensation and the theatre of flight', *Geografiska Annaler Series B* 90 (1): 29–47.

Adey, P. (2008b) 'Airports, mobility, and the calculative architecture of affective control', *Geoforum* 39 (1): 438–91.

Adey, P. (2009) 'Facing airport security: affect, biopolitics, and the preemptive securitisation of the mobile body', *Environment and Planning D – Society & Space* 27: 274–95.

Adey, P., Budd, L. and Hubbard, P. (2007) 'Flying lessons: exploring the social and cultural geographies of global air travel', *Progress in Human Geography* 31 (6): 773–91.

Agamben, G. (1998) *Homo sacer: sovereign power and bare life*. Stanford, Calif.: Stanford University Press.

Agamben, G. (2004) *The open: man and animal*. Stanford, Calif.: Stanford University Press.

Agre, P. (1994) 'Understanding the digital individual', *The Information Society* 10: 73–6.

Agre, P. (2001) 'Your face is not a bar code: arguments against automatic face recognition in public places', *Whole Earth* 106: 74–7.

Ahmed, S. (2004) *The cultural politics of emotion*. Edinburgh: Edinburgh University Press.

Air League of the British Empire (1923) *Facts about flying and the civil uses of aviation*. [London.]

Air Raid Precautions (1935) *Air Raid Precautions handbook no. 2: Anti-gas precautions and first aid for air raid casualties*. London: HMSO.

Air Raid Precautions (1938a) *Air Raid Precautions handbook no. 8: The duties of air raid wardens*. London: HMSO.

Air Raid Precautions (1938b) *The protection of your home against air raids*. London: HMSO.

Air Survey Committee (1923) *Report of the Air Survey Committee*. War Office. London: HMSO

Albrecht, C. and McIntyre, L. (2006) *Spychips*. London: Plume Penguin.

Alexander, L. (1945) *The treatment of shock from prolonged exposure to cold, especially in water*. Combined Intelligence Objectives Subcommittee, Item No. 24, File No. XXVI–37, 1–228.

Allen, J. (2006) 'Ambient power: Berlin's Potsdamer Platz and the seductive logic of public spaces', *Urban Studies* 43 (2): 441–55.

Amoore, L. (2006) 'Biometric borders: governing mobilities in the war on terror', *Political Geography* 25: 336–51.

Amoore, L. (2007) 'Vigilant visualities: the watchful politics of the war on terror', *Security Dialogue* 38: 215–32.

Amoore, L. (2009) 'Algorthmic war: everyday geographies of the war on terror', *Antipode* 41: 49–69.

Amoore, L. and de Goede, M. (2008) 'Transactions after 9/11: the banal face of the preemptive strike', *Transactions – Institute of British Geographers* 33 (2): 173–85.

Amoore, L. and Hall, A. (2009) 'Taking bodies apart: digitised dissection and the body at the border', *Environment and Planning D – Society & Space* 27 (3): 444–64.

Anderson, B. (2004) 'Time-stilled space-slowed: how boredom matters', *Geoforum* 35 (6): 739–54.

Anderson, B. (2006) 'Becoming and being hopeful: towards a theory of affect', *Environment and Planning D – Society & Space* 24 (5): 733–52.

Anderson, B. (2007) 'Hope for nanotechnology: anticipatory knowledge and the governance of affect', *Area* 39 (3): 156–65.

Anderson, B. (2010a in press) 'Effects-based airpower and the affective geographies of the war on terror', *Cultural Geographies*.

Anderson, B. (2010b in press) 'Population and affective perception: biopolitics and anticipatory action in US counterinsurgency doctrine', *Antipode*.

Anderson, B. and Harrison, P. (2006) 'Questioning affect and emotion', *Area* 38 (3): 333–5.

Andreas, P. (2003) 'Redrawing the line: borders and security in the twenty-first century', *International Security* 28 (2): 78–111.

Ash, E. (1999) 'Terror targeting: the morale of the story', *Aerospace Power Journal*, Winter: 33–47.

Association of Building (1938) *ARP … A report on the design, equipment and cost of air-raid shelters. Reprinted from the Architects' Journal … Second edition* London.

Augé, M. (1995) *Non-places: introduction to an anthropology of supermodernity*. London, New York: Verso.

Babington, A. (1997) *Shell shock: a history of changing attitudes to war neurosis*. London: Leo Cooper.

Baddington-Smith, C. (1958) *Evidence in camera: the story of photographic intelligence in World War II*. London: Chatto and Windus.

Balchin, W. G. V. (1947) *Air transport and geography*. London: Royal Geographical Society.

Balibar, É. (1991) 'Es Gibt Keinen Staat in Europa: Racism and Politics in Europe Today', *New Left Review* I (186): 5–19.

Balibar, É. (2004) *We, the people of Europe: reflections on transnational citizenship*. Princeton: Princeton University Press.

Balibar, É. and Wallerstein, I. (eds) (1991) *Race, nation, class: ambiguous identities*. London: Verso.

Banner, S. (2008) *Who owns the sky?* Cambridge, Mass.: Harvard University Press.

Barber, N. (2004) *The war of running dogs*. London: Cassell Phoenix.

Barksdae Maynard, W. (1999) '"An ideal life in the woods for boys": architecture and culture in the earliest summer camps', *Winterthur Portfolio* 34 (1): 3–29.

Bartlett, E. (2009) 'Profound psychological damage in Gaza', *Electronic Infitada*, 21 January (http://electronicintifada.net/v2/article10230.shtml).

Bartlett, F. C. (1943) 'Fatigue following highly skilled work', *Proceedings of the Royal Society London B* 131: 247–57.

Barton, G. (2001) 'Empire forestry and the origins of environmentalism', *Journal of Historical Geography* 27 (4): 529–52.

Bates, R. (2000) 'The spending state of mind', *Airport World* 5 (2): 44–6.f

Battersby, A. (1964) *Network analysis for planning and scheduling*. London: Macmillan.

Baxstrom, R. (2000) 'Governmentality, bio-power and the emergence of the Malayan-Tamil subject on the plantations of colonial Malaya', *Crossroads: An Interdisciplinary Journal of Southeast Asian Studies* 14 (2): 49–78.

Beazeley, G. A. (1919) 'Air photography in archaeology', *Geographical Journal* 53 (5): 330–5.

Bednarek, J. R. D. (2001) *America's airports: airfield development, 1918–1947*. College Station: Texas A&M University Press.

Bednarek, J. R. D. (2005) 'The flying machine in the garden: parks and airports, 1918–1938', *Technology and Culture* 46 (2): 350–73.

Behnke, P. (2005) 'Airport retailing: a key to financial performance', *Airport World* 9 (3): 11–15.

Bell, C. (2006) 'Surveillance strategies and populations at risk: biopolitical governance in Canada's national security policy', *International Political Sociology* 37 (2): 147–65.

Bennett, C. (2006) 'What happens when you book an airline ticket (revisited): the computer-assisted passenger profiling system and the globalization of personal data', in E. Zureik and M. B. Salter (eds), *Global surveillance and policing: borders, security, identity*. Cullompton: Willan.

Bennett, J. (2004) 'The force of things: steps towards an ecology of matter', *Political Theory* 32 (3): 347–72.

Benvenisti, M. (2004) *Abu Ghraib: the politics of torture*. Berkeley: North Atlantic Books.

Bernard, A. (2004) 'Lessons from Iraq and Bosnia on the theory and practice of no-fly zones', *Journal of Strategic Studies* 27 (3): 454–78.

Bialer, U. (1980) *The shadow of the bomber: the fear of air attack and British politics, 1932–1939*. London: Royal Historical Society.

Biddle, T. D. (2004) *Rhetoric and reality: the evolution of British and American ideas about strategic bombing, 1914–1945*. Princeton: Princeton University Press.

Biggs, D. (2006) 'Reclamation nations: the U.S. Bureau of Reclamation's role in water management and nation building in the Mekong Valley, 1945–1975', *Comparative Technology Transfer and Society* 4 (3): 225–46.

Bigo, D. (2006) 'Security, exception, ban and surveillance', in D. Lyon (ed.), *Theorizing surveillance: the Panopticon and beyond*. London: Willan.

Bingham, N. (2004) 'Arrivals and departures: civil airport architecture in Britain during the inter-war period', in J. Holder and S. Parissien (eds), *The architecture of British transport in the twentieth century*. New Haven, London: Yale University Press.

Bingham, P. T. (2000) 'Transforming warfare with effects-based joint operations', *Air and Space Power Journal*, October: 106–32.

Birley, J. (1920) 'The principles of medical science as applied to military aviation', delivered before the Royal College of Physicians of London, *The Lancet*, 12 June: 1147–51.

Bishop, R. and Clancey, G. (2004) 'The city-as-target, or perpetuation and death', in S. Graham (ed.), *Cities, war and terrorism*. Oxford: Blackwell.

Bishop, R. and Phillips, J. (2002a) 'Sighted weapons and modernist opacity: aesthetics, poetics, prosthetics', *boundary 2* 29 (2): 157–79.

Bishop, R. and Phillips, J. (2002b) 'Manufacturing emergencies', *Theory, Culture & Society* 19: 91–102.

Bishop, S., Grayling T. and Institute for Public Policy Research (2003) *The sky's the limit: policies for sustainable aviation*. London: Institute for Public Policy Research.

Bissell, D. (2008) 'Comfortable bodies: sedentary affects', *Environment and Planning A* 40 (7): 1697–712.

Bissell, D. (2009) 'Conceptualizing differently-mobile passengers: geographies of everyday encumbrance in the railway station', *Social and Cultural Geography* 10 (2): 173–95.

Blache, J. (1921) 'Modes of life in the Moroccan countryside: interpretations of aerial photographs', *Geographical Review* 11 (4): 477–502.

Blaut, J. (1997) 'Piagetian pessimism and the mapping abilities of young children: a rejoinder to Liben and Downs', *Annals of the Association of American Geographers* 87 (1): 168–77.

Blomley, N. K. (1994) *Law, space, and the geographies of power*. New York, London: Guilford.

Blomley, N. K. (2003) 'Law, property and the spaces of violence: the frontier, the survey, and the grid', *Annals of the Association of American Geographers* 93 (1): 121–41.

Boehmer, E. (2005) 'Introduction', in R. Baden-Powell (ed.), *Scouting for Boys*. Oxford: Oxford University Press.

Bogod, D. (2004) 'Editorial 1: the Nazi hypothermia experiments: forbidden data?', *Anaesthesia* 59: 1155–9.

Bohn, W. (2006) 'The poetics of flight: futurist "aeropoesia"', *MLN* 121: 207–24.

Bosher, L. S., Dainty, A., Carrillo, P. and Glass, J. (2007) 'Built-in resilience to disasters: a pre-emptive approach', *Engineering Construction and Architectural Management* 14 (5): 434–46.

Bourke, J. (1996) *Dismembering the male: men's bodies, Britain and the Great War*. London: Reaktion.

Bourke, J. (2005) *Fear: a cultural history*. London: Virago Press.

Bourne, R. (1928) *Aerial survey in relation to the economic development of new countries, with special reference to an investigation carried out in Northern Rhodesia*. Oxford: Clarendon Press.

Bourne, R. (1930) 'Air survey within the Empire', *Air Annual of the British Empire* 2: 15–18.

Bousquet, A. (2009) *The scientific way of warfare: order and chaos on the battlefields of modernity*. London, C. Hurst and Co. Publishers.

Boyer, C. (2003) 'Aviation and the aerial view: Le Corbusier's spatial transformations in the 1930s and 1940s', *Diacritics* 33 (3–4): 93–116.

Braaksma, J. (1976) 'Time-stamping: a new way to survey pedestrian traffic in airport terminals', *Transportation Research Record* 588: 27–34.

Braaksma, J. and Shortreed, J. H. (1971) 'Improving airport gate usage with critical path method', *ASCE Transportation Engineering Journal* 97: 187–203.

Bradstock, A. (2000) *Masculinity and spirituality in Victorian culture*. Basingstoke: Macmillan.

Braun, H.-J. (1995) 'The airport as symbol: air transport and politics at Berlin-Templehof, 1923–1948', in. W. M. Leary (ed.), *From airships to Airbus: the history of civil and commercial aviation, Vol. 1: Infrastructure and environment*. Washington, DC, London: Smithsonian Institution Press.

Brennan, T. (2003) *The transmission of affect*. Ithaca, NY, London: Cornell University Press.

Brett, L. (1962) 'Arrivals and departures', *Architectural Review* 18: 4–16.

Brinn, D. (2005) 'Israeli airport technology detects intent of terrorists', *Israel21c: A Focus Beyond*, 8 May (http://www.israel21c.com).

Brockway, L. (2002) *Science and colonial expansion: the role of the British Royal Botanic Gardens*. New Haven: Yale University Press.

Broeders, D. (2007) 'The new digital borders of Europe: EU databases and the surveillance of irregular migrants', *International Sociology* 22 (1): 71–92.

Brooker, P. (2003) 'Single sky and free market', *Institute of Economic Affairs*, Summer: 45–51.

Brophy, K. and Crowley, D. (eds) (2005) *From the air: understanding aerial archaeology*. Stroud: Tempus.

Brown, F. (1941) 'Civilian psychiatric air-raid casualties', *The Lancet*, 31 May: 686–91.

Bruno, G. (2002) *Atlas of emotion: journeys of art, architecture and film*. New York, London: Verso.

Bryant, R. (1997) *The political ecology of forestry in Burma, 1824–1994*. New York: C. Hurst and Co.

Budd, L. (2009) 'Air craft: producing UK airspace', in S. B. Cwerner, S. Kesselring and J. Urry (eds), *Aeromobilities*. London: Routledge.

Budd, L., Bell, M. and Brown, T. (2010 in press) 'Of plagues, planes and politics: controlling the global spread of infectious diseases by air', *Political Geography*.

Buhler, C. (1933) 'The child and its activity with practical material', *British Journal of Educational Psychology* 3: 27–41.

Bull, H., Kingsbury, B. and Roberts, A. (eds) (1992) *Hugo Grotius and international relations*. London: Clarendon Press.

Bull, M. (2007) 'Vectors of the biopolitical', *New Left Review* 45: 7–25.

Burt, J. (2006) 'Solly Zuckerman: the making of a primatological career in Britain, 1925–1945', *Studies in History and Philosophy of Science Part C* 37 (2): 295–310.

Butler, D. L. (2001) 'Technogeopolitics and the struggle for control of world air routes, 1910–1928', *Political Geography* 20 (5): 635–58.

Butler, J. (1999) *Gender trouble: feminism and the subversion of identity* New York; London, Routledge.

Calder, A. (1991) *The myth of the Blitz*. London: Pimlico.

Calder, R. (1940) 'Sleep we must', *New Statesman and Nation*, 14 September: 252–3.

Calder, R. (1999) 'Bernal at war', in B. Swann and F. Aprahamian (eds), *J. D. Bernal: a life in science and politics*. London: Verso.

Cambridge Scientists' Anti-War Group (1937) *The protection of the public from aerial attack: being a critical examination of the recommendations put forward by the Air Raid Precautions Department of the Home Office*. London: Victor Gollancz.

Campbell, D. (2007) 'Geopolitics and visuality: sighting the Darfur conflict', *Political Geography* 26: 357–82.

Campbell, D. (2009) 'Tele-vision: satellite images and security', *Source* 56: 16–23.

Canguilhem, G. (2008) *Knowledges of life*. New York: Fordham University Press.

Cantlie, J. (1906) *Physical efficiency: a review of the deleterious effects of town life upon the population of Britain, with suggestions for their arrest*. London: G. P. Putnam and Sons.

Caprotti, F. (2008) 'Technology and geographical imaginations: representing aviation in 1930s Italy', *Journal of Cultural Geography* 25(2): 181–205.

Carey, S. D. and Read, R. S. (2006) 'Five propositions regarding effects-based operations', *Air and Space Power Journal* 20 (1): 63–74.

Carruthers, S. (2005) 'Two faces of 1950s terrorism: the film presentation of Mau Mau and the Malayan Emergency', in J. D. Slocum (ed.), *Terrorism, media, liberation*. New York: Rutgers University Press.

Castells, M. (1996) *The rise of the network society*. Oxford: Blackwell.

Chambers, I. (1994) *Migrancy, culture, identity*. London, New York: Routledge.

Charles, M. B., Barnes, P., Ryan, N. and Clayton, J. (2007) 'Airport futures: towards a critique of the aerotropolis model', *Futures* 39: 1007–28.

Charlton, L. E. O. (1935) *War from the air, past, present, future*. London: T. Nelson & Sons.

Clarke, D. (2000) 'Scareships over Britain: the airship wave of 1909' (http://www.ufo.se).

Clarke, L. B. (1999) *Mission improbable: using fantasy documents to tame disaster*. Chicago: University of Chicago Press.

Clarke, L. B. (2006) *Worst cases: terror and catastrophe in the popular imagination*. Chicago: University of Chicago Press.

Coaffee, J., Moore, C., Fletcher, D. and Bosher, L. (2008) 'Resilient design for community safety and terror-resistant cities', *Municipal Engineer* 16 (1): 103–10.

Coates, J. (1992) *Suppressing insurgency: an analysis of the Malayan Emergency, 1948–1954*. Boulder, Colo., Oxford: Westview Press.

Cobham, A. J. (1978) *A time to fly*. London: Shepheard-Walwyn.

Cocroft, W., Thomas, R. J. C., Barnwell, P. S. and English, H. (2003) *Cold war: building for nuclear confrontation 1946–1989*. Swindon: English Heritage.

Cohen, E. (1949) *Human behaviour in the concentration camp*. London: Jonathan Cape.

Collier, P. (1994) 'Innovative military mapping using aerial photography in the First World War: Sinai, Palestine and Mesopotamia 1914–1919', *Cartographic Journal* 31 (22): 100–4.

Collier, P. (2002) 'The impact on topographic mapping of developments in land and air survey: 1900–1939', *Cartography and Geographic Information Science* 29 (3): 155–174.

Collier, P. (2006) 'The work of the British government's Air Survey Committee and its impact on mapping in the Second World War', *The Photogrammetric Record* 21 (114): 100–9.

Collier, P. and Inkpen, R. (2003) 'Photogrammetry in the Ordnance Survey from Close to Macleod', *The Photogrammetric Record* 18 (103): 224–43.

Collier, S. and Lakoff, A. (2008) 'Distributed preparedness: the spatial logic of domestic security in the United States', *Environment and Planning D – Society & Space* 26 (1): 7–28.

Collitt, R. (2008) 'Brazil bets on technology to control huge Amazon', Reuters, 15 October (http://www.reuters.com/article/latestCrisis/idUSN15321542).

Corn, J. J. (1983) *The winged gospel: America's romance with aviation, 1900–1950* New York: Oxford University Press.

Cote-Boucher, K. (2008) 'The diffuse border: intelligence-sharing, control and confinement along Canada's smart border', *Surveillance and Society* 5 (2): 142–65.

Coward, M. (2006) 'Against anthropocentrism: the destruction of the built environment as a distinct form of political violence', *Review of International Studies* 32: 419–37.

Craik, J. W. (1943) *The nature of explanation.* Cambridge: Cambridge University Press.

Craik, J.W. (1947) 'Theory of the human operator in control systems', *British Journal of Psychology* 38: 56–61.

Crandall, J. (2006) 'Precision + guided + seeing', *CTheory* (http://www.ctheory.net/articles.aspx?id=502).

Crandall, J. (2008) 'An actor prepares', *CTheory* (http://www.ctheory.net/articles.aspx?id=590).

Crang, M. (2002) 'Between places: producing hubs, flows, and networks', *Environment and Planning A* 34 (4): 569–74.

Crary, J. (1990) *Techniques of the observer: on vision and modernity in the nineteenth century.* Cambridge, Mass.: MIT Press.

Crary, J. (1999) *Suspensions of perception: attention, spectacle, and modern culture.* Cambridge, Mass.: MIT Press.

Cresswell, T. (2001) 'The production of mobilities', *New Formations* 43 (1): 11–25.

Cresswell, T. (2006a) *On the move: the politics of mobility in the modern West.* London: Routledge.

Cresswell, T. (2006b) '"You cannot shake that shimmie here": producing mobility on the dance floor', *Cultural Geographies* 13 (1): 55–77.

Crichton-Miller, H. (1941) 'Somatic factors conditioning air-raid reactions', *The Lancet*, 12 July: 31–4.

Cronin, M. (2007) 'Northern visions: aerial surveying and the Canadian mining industry, 1919–1928', *Technology and Culture* 48 (April): 303–30.

Cronon, W., Miles, G. and Gitlin, J. (1992) *Under an open sky: rethinking America's Western past.* New York, London: W. W. Norton.

Crosby, T. L. (1986) *The impact of civilian evacuation in the Second World War.* London: Croom Helm.

Crowley, D. and Pavitt, J. (2008) *Cold war modern: design 1945–1970.* London: V&A.

Cruddas, C. (2003) *Those fabulous flying years: joy-riding and flying circuses between the wars.* Tunbridge Wells: Air Britain.

Cunningham, V. (1988) *British writers of the thirties.* Oxford: Oxford University Press.

Currell, S. and Cogdell, C. (2006) *Popular eugenics: national efficiency and American mass culture in the 1930s.* Athens, Ohio: Ohio University Press.

Curry, M. (2004) 'The profiler's question and the treacherous traveller: narratives of belonging in commercial aviation', *Surveillance and Society* 1 (4): 475–99.

Curwen, E. C. (1929) *Air photography and economic history: the evolution of the corn-field.* London: Economic History Society.

Cwerner, S. B. (2009) 'Introducing aeromobilities', in S. B. Cwerner, S. Kesselring and J. Urry (eds), *Aeromobilities.* London: Routledge.

Cwerner, S. B., Kesslering S. and Urry J. (2009) *Aeromobilities* London, Routledge.

Dalloul, M. (2009) 'Gaza voices: "People still afraid"', Al Jazeera, 18 January (http://english.aljazeera.net/news/middleeast/2009/01/200911872210319480.html).

Damasio, A. R. (2003) *Looking for Spinoza: joy, sorrow and the feeling brain.* London: Heinemann.

Daniels, S. and Rycroft, S. (1993) 'Mapping the modern city: Alan Sillitoe's Nottingham novels', *Transactions of the Institute of British Geographers* 18 (4): 460.

Darwin, C. (1892) *The expression of the emotions in man and animals.* London: John Murray.

Davies, J. L. (1939) *Air raid precautions: what to do in emergency.* London: George Newnes.

Davis, B. L. and McGregor, M. (2007) *The German home front 1939–45.* Oxford: Osprey.

Davis, M. (2002) *Dead cities, and other tales.* New York, New Press; London: I. B. Tauris.

Davis, M. (2009) *City of slums.* New York: Verso.

Davis, T. (2002) 'Between history and event: rehearsing nuclear war survival', *The Drama Review* 46 (4): 11–50.

Dawbarn, G. (1932) *A brief report on the design and construction of civil airports in the United States of America.* Royal Institute of British Architects Aerodrome Committee, London (RAF Archives).

de Botton, A. (2002) *The art of travel.* London: Pantheon.

de Certeau, M. (1984) *The practice of everyday life.* Berkeley: University of California Press.

de Costa, G. (2001) 'Brazil's SIVAM: as it monitors the Amazon, will it fulfill its human security promise?', *ESCP Report* 7: 47–58

Deery, P. (2007) 'Malaya, 1948: Britain's Asian cold war?', *Journal of Cold War Studies* 9 (1): 29–54.

Delanty, G. (2006) 'Borders in a changing Europe: dynamics of openness and closure', *Comparitive European Politics* 4: 183–202.

Deleuze, G. (1988) 'Ethology: Spinoza and us', in J. Crary and S. Kwinter (eds), *Incorporations.* New York: Zone Books.

Deleuze, G. and Guattari, F. (1988) *A thousand plateaus: capitalism and schizophrenia.* London: Athlone Press.

Demeritt, D. (2001) 'Scientific forest conservation and the statistical picturing of nature's limits in the Progressive-era United States', *Environment and Planning D – Society & Space* 19 (4): 431–59.

Demetz, P. (2002) *The air show at Brescia, 1909.* New York: Farrar, Straus and Giroux.

Dennis, K. and Urry, J. (2009) *After the car.* Cambridge: Polity.

Dent, M. (2002) 'Staging disaster reporting live (sort of) from Seattle', *The Drama Review* 48 (4): 109–34.

DePaulo, B. M., Lindsay, J. J., Malone, B. E., Muhlenbruck, L., Charlton, K. and Cooper, H. (2003) 'Cues to deception', *Psychological Bulletin* 129 (1): 74–112.

Der Derian, J. (2001) *Virtuous war: mapping the military–industrial–media–entertainment network.* London: Routledge.

Deriu, D. (2006) 'The ascent of the modern *planeur*: aerial images and urban imaginary in the 1920s', in C. Emden, C. Keen and D. Midgley (eds), *Imagining the city, vol. 1: The art of urban living.* Oxford: Peter Lang.

Derudder, B., Witlox, F. and Faulconbridge, J. (2008a) 'Airline networks and urban systems', *Geojournal* 71 (1): 1–3.

Derudder, B., Witlox, F., Faulconbridge, J. and Beaverstock, J. V. (2008b) 'Airline data for global city network research: reviewing and refining existing approaches', *Geojournal* 71 (1): 5–18.

Dierikx, M. (1991) 'Struggle for prominence: clashing Dutch and British interests on the colonial air routes, 1918–1942', *Journal of Contemporary History* 26: 333–51.

Diken, B. and Laustsen C. B. (2006) 'The camp', *Geografiska Annaler Series B* 88 (4): 443–52.

Dillon, M. (2003) 'Virtual security: a life science of (dis)order', *Millennium – Journal of International Studies* 32 (3): 531–58.

Dillon, M. (2004) 'Correlating sovereign and biopower', in J. Edkins, V. Pin-Fat and M. J. Shapiro (eds), *Sovereign lives.* London: Routledge.

Dillon, M. (2007a) 'Governing through contingency: the security of biopolitical governance', *Political Geography* 26 (1): 41–7.

Dillon, M. (2007b) 'Governing terror: the state of emergency of biopolitical governance', *International Political Sociology* 1: 7–28.

Dillon, M. and Lobo-Guerrero, L. (2008) 'Biopolitics of security in the 21st century', *The Review of International Studies* 34: 265–92.

Dillon, M. and Lobo-Guerrero, L. (2009) 'The biopolitical imaginary of species-being', *Theory, Culture & Society* 26 (1): 1–23.

Dillon, M. and Reid, J. (2009) *The liberal way of war.* London: Routledge.

Dival, C. and Revill, G. (2005) 'Cultures of transport: representation, practice, technology', *Journal of Transport Geography* 26 (1): 99–112.

Dixon, D. (2007) 'A benevolent and sceptical inquiry: exploring "Fortean geographies" with the Mothman', *Cultural Geographies* 14 (2): 189–210.

Dobbs, B. (1973) *Edwardians at play: sport, 1890–1914.* London: Pelham.

Dodge, M. and Kitchin, R. (2004) 'Flying through code/space: the real virtuality of air-travel', *Environment and Planning A* 36 (2): 195–211.

Dodge, M. and Kitchin R. (2005) 'From A to B but only if C: air travel and the banal geographies of control creep', Association of American Geographers Annual Conference, Denver.

Donne, M. (1988) 'Consumers' view of world air transport', in *Airports for the people: proceedings of the Eighth World Airports Conference*, Institution of Civil Engineers, London, 2–5 June 1987.

Douglas, A. (1999) 'Skyscrapers, airplanes, and airmindedness: "the necessary angel"', in R. G. O'Meally (ed.), *The jazz cadence of American culture*. New York: Columbia University Press.

Douthwaite, A. H. and Leak, W. N. (1940) 'Safeguarding sleep', *The Lancet*, 2 November: 568.

Dowling, A. (2001) *Manliness and the male novelist in Victorian literature*. Aldershot: Ashgate.

Dowson, E. (1921) 'Notes on aeroplane photography in the Far East', *The Geographical Journal* 58 (5): 359–70.

Durkheim, É. (1997) *Suicide*. London: Free Press.

Edensor, T. (2001) 'Performing tourism, staging tourism: (re)producing tourist space and practice', *Tourist Studies* 1 (1): 59.

Edensor, T. (2002) *National identity, popular culture and everyday life*. Oxford: Berg.

Edgarton, D. (1991) *England and the aeroplane: an essay on a militant and technological nation*. Basingstoke, Macmillan in association with the Centre for the History of Science, Technology and Medicine, University of Manchester.

Ekman, P. (2001) *Telling lies: clues to deceit in the marketplace, politics, and marriage*. New York, London: W. W. Norton.

Ekman, P. (2003) *Emotions revealed: understanding faces and feelings*. London: Weidenfeld & Nicolson.

Ekman, P. (2006) *Darwin and facial expression: a century of research in review*. Cambridge, Mass.: Malor Books.

Ekman, P., O'Sullivan, M. and Frank, M. G. (1999) 'A few can catch a liar', *Psychological Science* 10: 263–6.

Elden, S. (2005) 'Territorial integrity and the war on terror', *Environment and Planning A* 37: 2083–104.

Elden, S. (2007) 'Terror and territory', *Antipode* 39 (5): 821–45.

Elden, S. and Williams A. (2009) 'The territorial integrity of Iraq, 2003–2007', *Geoforum* 40 (3): 407–17.

Elmer, G. (2004) *Profiling machines: mapping the personal information economy*. Cambridge, Mass.: MIT Press.

Epstein, C. (2007) 'Guilty bodies, productive bodies, destructive bodies: crossing the biometric borders', *International Political Sociology* 1 (2): 149–64.

Ericson, R. V. (2006) 'Ten uncertainties of risk-management approaches to security', *Canadian Journal of Criminology and Criminal Justice* 48 (3): 345–57.

Fauchille, P. (1901) 'Le domain aerien et le régime juridique des aerostats', *Revue générale de droit international public* 8: 414–85.

Feldman, L. C. (2007) 'Terminal exceptions: law and sovereignty at the airport threshold', *Law, Culture and the Humanities* 3 (2): 320–44.

Felker, W. (1998) 'Airpower, chaos, and infrastructure: lord of the rings', *Maxwell Paper* 14.

Finnegan, T. (2006) *Shooting the Front: Allied aerial reconnaissance and photographic interpretation on the Western Front – World War I*. Washington, DC: Nation Defence Intelligence College.

Flack, M. (1919) 'Some simple tests of physical efficiency', *The Lancet*, 8 February: 210–12.

Flack, M. (1920) 'The medical requirements for air navigation', *The Lancet*, 23 October: 838–42.

Flack, M. (1921) 'The Milroy Lectures: respiratory efficiency in relation to health and disease', *The Lancet*, 1 October: 693–6.

Follows, R. (1999) *The jungle beat*. London: Eye Books.

Forster, J. (2003) 'Ludendorff and Hitler in perspective: the battle for the German soldier's mind, 1917–1944', *War in History* 10 (3): 321–34.

Fort, C. (1974) *The complete works of Charles Fort*. New York: Dover.

Foucault, M. (1977) *Discipline and punish: the birth of the prison*. London: Allen Lane.

Foucault, M. (1980) *Power/knowledge: selected interviews and other writings, 1972–1977*. New York: Harvester; Harlow: Longman.

Foucault, M. (2003) *Society must be defended*. New York, Picador.

Foucault, M. (2007) *Security, territory, population: lectures at the Collège de France, 1977–78*. Basingstoke: Palgrave Macmillan.

Foulkes, C. H. (1939) *Commonsense and ARP: a practical guide for householders and business managers*. London: C. A. Pearson.

Fox, W. L. (2009) *Aereality: on the world from above*. New York: Counterpoint.

Frank, M. G. and Ekman, P. (1997) 'The ability to detect deceit generalizes across different types of high-stake lies', *Journal of Personality and Social Psychology* 72 (6): 1429–39.

Friedberg, A. (1993) *Window shopping: cinema and the postmodern*. Berkeley: University of California Press.

Friedrich, J. (2006) *The fire: the bombing of Germany 1940–1945*. New York: Columbia University Press.

Fritzsche, P. (1992) *A nation of fliers: German aviation and the popular imagination*. Cambridge, Mass.: Harvard University Press.

Fritzsche, P. (1993) Machine dreams: airmindedness and the reinvention of Germany. *The American Historical Review* 98: 685–709.

Fuller, G. (2002) 'The arrow-directional semiotics: wayfinding in transit', *Social Semiotics* 12 (3): 231–44.

Fuller, G. and Harley, R. (2004) *Aviopolis: a book about airports*. London: Blackdog.

Gagen, E. (2004) 'Making America flesh: physicality and nationhood in early twentieth-century physical education reform', *Cultural Geographies* 11 (4): 417–42.

Gagen, E. (2006) 'Measuring the soul: psychological technologies and the production of physical health in Progressive Era America', *Environment and Planning D – Society & Space* 24 (6): 827–50.

Galison, P. (1994) 'The ontology of the enemy: Norbert Wiener and the cybernetic vision', *Critical Inquiry* 21 (1): 228–66.

Galison, P. (2001) 'War against the center', *Grey Room* 4: 6–33.

Garlasco, M. (2009) 'Remote control death', *Human Rights Watch*, 20 March (http://www.hrw.org/en/news/2009/03/20/remote-control-death).

Gaskin, M. J. (2005) *Blitz: the story of 29th December 1940*. London: Faber and Faber.

Gates, K. (2006) 'Identifying the 9/11 "faces of terror": the promise and problem of facial recognition technology', *Cultural Studies* 20 (4–5): 417–40.

Gavish, D. (2005) *A survey of Palestine under the British Mandate, 1920–1948*. London: Routledge.

Gaza.blogspot (2006) A Mother from Gaza, 'Return of the drones' (http://a-mother-from-gaza.blogspot.com/2006/01/return-of-drones.html).

Gibson, J. W. (2000) *The perfect war: technowar in Vietnam*. New York: Avalon.

Gibson, T. M. and Harrison, M. (2005) 'Aviation medicine in the United Kingdom: from the End of World War I to the end of World War II, 1919–1945', *Aviation, Space and Environmental Medicine* 76 (6): 689–91.

Gilboy, J. (1997) 'Implications of "third party" involvement in enforcement: the INS, illegal travelers, and international relations', *Law and Society Review* 31 (3): 505–30.

Gilcraft, R. W. B. (1927) *Training in tracking*. London: Pearson.

Gilcraft, R. W. B. (1932) *Wolf cubs*. London: Pearson.

Gilcraft, R. W. B. (1934) *Wide games*. London: Pearson.

Gladwell, M. (2005) *Blink: the power of thinking without thinking*. London: Allen Lane.

Goldstein, L. (1986) *The flying machine and modern literature*. Bloomington: Indiana University Press.

Gollin, A. (1989) *No longer an island: the impact of air power on the British people and their government, 1909–14*. Basingstoke: Macmillan.

Gordon, A. (2004) *Naked airport: a cultural history of the world's most revolutionary structure*. New York: H. Holt, Metropolitan Books.

Gottdiener, M. (2000) *Life in the air: surviving the new culture of air travel*. Lanham, Md: Rowman & Littlefield.

Graham, B. J. (1995) *Geography and air transport*. Chichester: John Wiley.

Graham, S. (2003) 'Lessons in urbicide', *New Left Review* 19: 63–78.

Graham, S. (2004) 'Vertical geopolitics: Baghdad and after', *Antipode* 36 (1): 12–23.

Graham, S. (2005) 'Switching cities off: urban infrastructures and US air power', *City* 9: 169–94.

Graham, S. (2006) 'Urban metabolism as target: contemporary war as forced demodernization', in N. C. Heynen, M. Kaika and E. Swyngedouw (eds), *In the nature of cities: urban political ecology and the politics of urban metabolism*. London: Routledge.

Graham, S. (2008) 'War play: practising urban annihilation', in F. von Borries, S. Walz and M. Botther (eds), *Space time play: on the synergy between computer games, architecture, and urbanism*. Basel: Birkhauser.

Graham, S. (2010 in press) 'Interrupting the algorithmic gaze? Urban warfare and US military technology', in F. MacDonald (ed), *Observant states: geopolitics and visual culture*. London: I.B. Tauris.

Graham, S. and Marvin, S. (2001) *Splintering urbanism: networked infrastructures, technological mobilities and the urban condition*. London: Routledge.

Grayling, A. C. (2006) *Among the dead cities: was the Allied bombing of civilians in World War II a necessity or a crime?* London: Bloomsbury.

Gregory, D. (2004) *The colonial present: Afghanistan, Palestine, Iraq*. London: Routledge.

Gregory, D. (2007) 'In another time-zone, the bombs fall unsafely …: targets, civilians and late modern war', *Arab World Geographer* 9: 88–112.

Gregory, D. (2008) 'The biopolitics of Baghdad: counterinsurgency in the counter-city', *Human Geography* 1 (http://www.hugeog.com).

Gregory, D. (2010 in press) 'Doors into nowhere: dead cities and the natural history of destruction', in P. Meusburger, M. Heffernan and E. Wunder (eds), *Cultural memory*. Berlin: Springer.

Grey, S. (2006) *Ghost plane: the untold story of the CIA's secret rendition programme*. New York, C. Hurst and Publishers.

Griffin, S. (1993) *A chorus of stones: the private life of war*. New York: Anchor.

Grosscup, B. (2006) *Strategic terror: the politics and ethics of aerial bombardment*. Selangor, Malaysia: SIRD; London: Zed.

Grove, C. (1970) 'The airship wave of 1909', *Flying Saucer Review* 6 (6): 9–11.

Guiraudon, V. and Lahav, G. (2000) 'A reappraisal of the state sovereignty debate', *Comparative Political Studies* 33 (2): 163–95.

Hables-Gray, C. (1997) *Postmodern war: the new politics of conflict*. London: Routledge.

Hadfield, G. (1941) 'Discussion on the problem of blast injuries', *Proceedings of the Royal Society of Medicine* 34: 189–91.

Haggerty, K. D. & Ericson, R. V. (2000) 'The surveillant assemblage', *British Journal of Sociology* 51: 605–22.

Haldane, J. B. S. (1925) *Callinicus: a defence of chemical warfare*. London: Kegan Paul & Co.

Haley, B. (1978) *The healthy body and Victorian culture*. Cambridge, Mass., London: Harvard University Press.

Hall, D. E. (1994) *Muscular Christianity: embodying the Victorian age*. Cambridge: Cambridge University Press.

Hannam, K., Sheller, M. and Urry, J. (2006) 'Editorial: mobilities, immobilities and moorings', *Mobilities* 1 (1): 1–22.

Haraway, D. J. (1991) *Simians, cyborgs and women: the reinvention of nature*. London: Free Association Books.

Harcombe, J. C. (1946) *Gliding*. London: Air League of the British Empire.

Harley, R. (2009) 'Light-air-portals: visual notes on differential mobility', *media-culture journal* 12 (1) (http://journal.media-culture.org.au/index.php/mcjournal/).

Harris, R. and Paxman, J. (2002) *A higher form of killing: the secret history of gas and germ warfare*. London: Arrow.

Harrison, P. (2000) 'Making sense: embodiment and the sensibilities of the every-day', *Environment and Planning D – Society & Space* 18 (4): 497–518.

Harrison, P. (2010 in press) 'In the absence of practice', *Environment and Planning D – Society & Space*.

Harrisson, T. (2000) 'Lionel Wigram, battle drill and the British Army in the Second World War', *War in History* 7: 442–62.

Harrisson, T. (1941) 'Obscure nervous effects of air raids', *British Medical Journal* 1: 573–4.

Harrisson, T. (1979) *Living through the Blitz*. Harmondsworth: Penguin.

Harvey, D. (1989) *The condition of postmodernity*. Oxford: Blackwell.

Haslam, E. B. (1982) *The history of Royal Air Force Cranwell*. London: HMSO.

Hastings, M. (1979) *Bomber command*. New York: Dial Press.

Hawley, K. (2007) 'Beyond the checkpoint', *Leadership Journal*, 11 October (http://www.tsa.gov/press/journal/beyond_checkpoint.shtm).

Hayward, R. (2001) 'Our friends electric: mechanical models of mind in post-war Britain', in G. Bunn, D. Lovie and R. Richards (eds), *Psychology in Britain: historical essays and personal reflections*. London, Routledge.

Helbing, D., Keltsch, J. and Molnar, P. (1997) 'Modelling the evolution of human trail systems', *Nature* 6637: 47–9.

Helbing, D., Buzna, L., Johansson, A. and Werner, T. (2005) 'Self-organized pedestrian crowd dynamics: experiments, simulations, and design solutions', *Transportation Science* 39 (1): 1–24.

Hell, J. (2004) 'The angel's enigmatic eyes, or the gothic beauty of catastrophic history in W. G. Sebald's "Air war and literature"', *Criticism* 46 (3): 361–92.

Hewitt, K. (1983) 'Place annihilation – area bombing and the fate of urban places', *Annals of the Association of American Geographers* 73 (2): 257–84.

Hoag, H. and Ohman, M.-B. (2008) 'Turning water into power: debates over the development of Tanzania's Rufiji River Basin, 1945–1985', *Technology and Culture* 49 (July): 624–51.

Hofer, F. and Schwaniger, A. (2005) 'Using threat image projection data for assessing individual screener performance', *WIT Transactions on the Built Environment* 82: 417–26.

Holden, P. and Ruppel, R. R. (2003) *Imperial desire: dissident sexualities and colonial literature*. Minneapolis, London: University of Minnesota Press.

Holden, W. (1998) *Shell shock*. London: Channel 4 Books.

Holloway, J. and Kneale, J. (2008) 'Locating haunting: a ghost-hunter's guide', *Cultural Geographies* 15 (3): 297–312.

Home Office (2007) *Securing the UK border: our vision and strategy for the future* (http://www.ukba.homeoffice.gov.uk/sitecontent/documents/managingourborders/securingtheukborder/securingtheukborder.pdf).

Hookway, B. (2006) 'Cockpit', in B. Colomina, A. Brennan and J. Kim (eds), *Cold war hothouses*. Princeton: Princeton Architectural Press.

Hosmer, S. (1996) *Psychological effects of US air operations in four wars, 1941–1991: lessons for US commanders*. MR-576-AF. Santa Monica, Calif.: RAND Corporation.

Hotine, M. (1927) *The stereoscopic examination of air photographs*. Professional Papers of the Air Survey Committee No. 4. War Office, London: HMSO.

Huss, J. (1999) 'Exploiting the psychological effects of airpower: a guide for the operational commander', *Aerospace and Power Journal*, Winter: 23–32.

Hyde, A. (2002) *The first blitz: the German air campaign against Britain in the First World War*. Barnsley: Leo Cooper.

Hynes, S. (1968) *The Edwardian turn of mind*. London: Pimlico.

Ignatieff, D. (2001) *Virtual war: Kosovo and beyond*. London: Picador.

Ingold, T. (2005) 'The eye of the storm: visual perception and the weather', *Visual Studies* 20 (2): 97–104.

Introna, L. and Wood, D. (2005) 'Picturing algorithmic surveillance', *Surveillance and Society* 2 (2/3): 177–98.

Isin, E. (2007) 'Governing affects', paper presented at the Biopolitics of Security Workshop, Keele University, Keele, November.

Iyer, P. (2000) *The global soul: jet lag, shopping malls and the search for home*. London: Bloomsbury.

James, B. J. and Beaumont, R. A. (1971) 'The law of military plumage', *Transition* 39: 24–7.

Jeal, T. (1989) *Baden-Powell*. London: Century.

Jensen, O. B. and Richardson, T. D. (2004) *Making European space: mobility, power and territorial identity*. London: Routledge.

Jones, E. T. (2004) 'War and the practice of psychotherapy: the UK experience, 1939–1960', *Medical History* 48: 493–510.

Jones, E. T. and Wessely S. (2005) *Shell shock to PTSD: military psychiatry from 1900 to the Gulf War*. Maudsley Monograph. Hove: Psychology Press.

Jones, E. T., Woolven, R., Durodie, B. and Wessely, S. (2004) 'Civilian morale during the Second World War: responses to air raids re-examined', *Social History of Medicine* 17 (3): 463–79.

Jones, E. T., Woolven, R., Durodie, B. and Wessely, S. (2006) 'Public panic and morale: Second World War civilian responses reexamined in the light of the current anti-terrorist campaign', *Journal of Risk Research* 9 (1): 57–73.

Joseph, N. and Alex, N. (1972) 'The uniform: a sociological perspective', *American Journal of Sociology* 77 (4): 719–30.

Jouas, J.-M. (1998) 'No-fly zones: an effective use of airpower, or just a lot of noise', unpublished research paper. Cambrige, Mass.: Weatherhead Center for International Affairs.

Jung, C. (1979) *Flying saucers: a modern myth of things seen in the skies*. Princeton: Princeton University Press.

Kam, E. (2004) *Surprise attack: the victim's perspective*. Cambrige, Mass.: Harvard University Press.

Kapka, M. (2009) 'Open the windows', Gaza Community Mental Health Programme (http://www.gcmhp.net/File_files/openthewindow.html).

Kaplan, C. (1996) *Questions of travel: postmodern discourses of displacement*. Durham, NC, London: Duke University Press.

Kaplan, C. (2006a) 'Mobility and war: the cosmic view of US "air power"', *Environment and Planning A* 38 (2): 395–407.

Kaplan, C. (2006b) 'Precision targets: GPS and the militarisation of US consumer identity', *American Quarterly* 3: 693–713.

Kaplan, C. and Kelley, C. (2006) 'Dead reckoning: aerial perception and the social construction of targets', *Vectors Journal* 2 (2) (http://www.vectorsjournal.org/).

Karmi, O. (2006) 'Gaza in the vice', Middle East Report Online, 11 July (http://merip.org/mero//mero071106.html).

Kasarda, J. D. (1991a) 'An industrial/aviation complex for the future', *Urban Land*, August: 16–20.

Kasarda, J. D. (1991b) 'The fifth wave: the air cargo–industrial complex', *Portfolio: A Quarterly Review of Trade and Transportation* 4 (1): 2–10.

Kearns, G. (1984) 'Closed space and political practice: Fredrick Jackson Turner and Halford Mackinder', *Environment and Planning D – Society & Space* 21: 23–34.

Kendall, J. (1938) *Breathe freely! The truth about poison gas*. London, G. Bell.

Kennedy, S. (2000) 'The Croix de Feu, the Parti Social Français, and the politics of Aviation, 1931–1939', *French Historical Studies* 23 (2): 373–99.

Kern, S. (1983) *The culture of time and space, 1880–1918*. Cambridge, Mass.: Harvard University Press.

Kerner, B. S. (1999) 'Phase transitions in traffic flow', paper presented at conference *Social, traffic, and granular dynamics: traffic and granular flow '99*, Stuttgart, Germany.

Kesby, A. (2006) 'The shifting and multiple border and international law', *Oxford Journal of Legal Studies* 27 (1): 101–19.

Kesselring, S. (2008) 'Global transfer points: the making of airports in the mobile risk society', in S. B. Cwerner, S. Kesselring and J. Urry (eds), *Aeromobilities*. London: Routledge.

Kirby, L. (1997) *Parallel tracks: the railroad and silent cinema*. Durham, NC: Duke University Press.

Kirk, D. and Twigg, K. (1995) 'Civilising Australian bodies: the games ethic and sport in Victorian government schools, 1904–1945', *Sporting Traditions* 11 (2): 3.

Klauser, F. (2009) 'Interacting forms of expertise in security governance: the example of CCTV surveillance at Geneva International Airport;, *British Journal of Sociology* 60: 279–97.

Klock, A. (1999) 'Of cyborg technologies and fascistized mermaids: Guiannina Censi's *Aerodanze* in 1930s Italy', *Theatre Journal* 51 (4): 395–415.

Knowles, M. and Malmkjær, K. (1996) *Language and control in children's literature*. London, New York: Routledge.

Knowles, R. D., Shaw, J. and Docherty, I. (2007) *Transport geographies: mobilities, flows and spaces*. Oxford: Blackwell.

Kormer, R. (1972) *The Malayan Emergency in retrospect: organisation of a successful counterinsurgency effort*. R-957-ARPA. Santa Monica, Calif.: RAND Corporation.

Kuper, H. (1973) 'Costume and identity', *Comparative Studies in Society and History* 15 (3): 348–67.

Lakoff, A. (2007) 'Preparing for the next emergency', *Public Culture* 19: 247–71.

Lambert, A. P. N. (1995) *The psychology of air power*. London: Royal United Services Institute.

Lancet (1917a) 'Air-raid psychology and air-raid perils', *The Lancet*, 6 October: 540–1.

Lancet (1917b) 'Air-raid psychology', *The Lancet*, 14 July: 55–6.

Lancet (1918) 'The doctor and the airman', *The Lancet*, 16 March: 411.

Lancet (1920a) 'Medical requirements for air navigation', *The Lancet*, 23 October: 838–41.

Lancet (1920b) 'The medical examination of the civilian aeronaut', *The Lancet*, 8 May: 1026–7.

Lancet (1938a) 'Gas attacks and panic', *The Lancet*, 4 June: 1285–6.

Lancet (1938b) 'Air raids, discipline, and panic', *The Lancet*, 7 May: 1061.

Lancet (1940) 'Reactions to air-raid warnings', *The Lancet*, 10 August: 177.

Langeweische, W. (1998) *Inside the sky: a meditation on Flight*. New York: Pantheon Books.

Lassen, C. (2006) 'Aeromobility and work', *Environment and Planning A* 38 (2): 301–12.

Lassen, C., Smink, C. and Smidt-Jensen, S. (2009) 'Experience spaces, (aero)mobilities and environmental impacts', *European Planning Studies* 17 (6): 887–903.

Laurier, E. (2004) 'Doing office work on the motorway', *Theory, Culture & Society* 21: 261–73.

Lawrence, T. E. (1936) *The Mint*. London: Jonathan Cape.

Le Bon, G. (1925) *The crowd*. London: T. F. Unwin.

Le Corbusier (1935) *Aircraft*. London: Studio Publications.

Le Corbusier (1947) *The Four Routes*. London: Dennis Dobston.

Lee, W. (1920) 'Airplanes and geography', *Geographical Review* 10 (5): 310–25.

Leese, P. (2002) *Shell shock: traumatic neurosis and the British soldiers of the First World War*. New York: Palgrave.

Lefebvre, H. (1991) *The production of space*. Oxford, Cambridge, Mass.: Blackwell.

Leppard, D. (2007) 'Britain under attack as bombers strike at airport;, *The Sunday Times*, 1 July, Times Online (http://www.timesonline.co.uk/tol/news/uk/crime/article2010062.ece).

Levy, G. (2005) 'Demons in the skies of the Gaza Strip', *Haaretz*, 6 November (http://www.haaretz.com/hasen/spages/641839.html).

Lewis, C. A. (1936) *Sagittarius rising*. London: Peter Davies.

Lewis, N. (2000) 'The climbing body, nature and the experience of modernity', *Body and Society* 6 (3–4): 58–80.

Lewy, G. (1978) *America in Vietnam*. New York: Oxford University Press.

Life Magazine (1944) 'Mechanical brains: working in metal boxes, computing devices aim guns and bombs with inhuman accuracy', *Life Magazine*, 24 January: 66–72.

Lindholm, C. (1990) *Charisma*. Oxford: Blackwell.

Lindquist, S. (2000) *A History of Bombing*. New York: Granta.

Lipton, E. (2006) 'Faces, too, are searched at US airports', *New York Times*, 17 August.

Lisle, D. (2003) 'Site specific: medi(t)ations at the airport', in F. Debris and C. Weber (eds), *Rituals of mediation: international politics and social meaning*. London, Minneapolis: University of Minnesota Press.

Lobo-Guerrero, L. (2008) 'Pirates, stewards, and the securitisation of global circulation', *International Political Sociology* 2 (3): 219–35.

Lockyer, K. G. (1969) *An introduction to critical path analysis*. London: Pitman.

Lofgren, O. (1999) *On holiday: a history of vacationing*. Berkeley, London: University of California Press.

Lopez, B. (1999) *About this life: journeys on the threshold of memory*. London: Vintage.

Lorimer, H. (2005) 'Cultural geography: the busyness of being "more-than-representational"', *Progress in Human Geography* 29 (1): 83–94.

Lorimer, J. (2007) 'Nonhuman charisma', *Environment and Planning D – Society & Space* 24 (5): 911–32.

Lowe, K. (2008) *Inferno: the devastation of Hamburg 1943*. London: Penguin.

Luckhurst, R. (2002) 'The contemporary London gothic and the limits of the "spectral turn"', *Textual Practice* 16 (3): 526–45.

Lyon, D. (2003) *Surveillance after September 11*. Cambridge, Malden, Mass.: Polity.

Lyon, D. (2007) *Surveillance studies: an overview*. Cambridge, Malden, Mass.: Polity.

McConnell, A. and Drennan, L. (2006) 'Mission impossible? Planning preparing for crisis', *Journal of Contingencies and Crisis Management* 14 (2): 59–70.

McCormack, D. P. (2003) 'An event of geographical ethics in spaces of affect', *Transactions – Institute of British Geographers* 28 (4): 488–507.

McCormack, D. P. (2006) 'For the love of pipes and cables: a response to Deborah Thien', *Area* 38 (3): 330–2.

McCormack, D. P. (2008) 'Engineering affective atmospheres on the moving geographies of the 1897 Andree Expedition', *Cultural Geographies* 15 (4): 413–30.

McDonagh, K. (2008) 'Governing global security in the departure lounge', *Journal of Global Change and Governance* 1 (3): 1–19.

MacDonald, R. H. (1993) *Sons of the Empire: the frontier and the Boy Scout movement, 1890–1918.* Toronto, London: University of Toronto Press.

Macintyre, D. (2005) 'Palestinians "terrorised" by sonic boom flights', Palestinian Centre for Human Rights, 3 November (http://www.pchrgaza.ps/Library/donald.htm).

McKay, D. (2004) *Half the battle.* Manchester: Manchester University Press.

Macleod, M. N. (1919) 'Mapping from air photographs', *The Geographical Journal* 53 (6): 382–96.

McNeill, D. (2009) 'The airport hotel as business space', *Geografiska Annaller B* 91 (3): 219–28.

McNeill, W. H. (1995) *Keeping together in time: dance and drill in human history* Cambridge, Mass.: Harvard University Press.

Mangan, J. A. (1998) *The games ethic and imperialism: aspects of the diffusion of an ideal.* London: Frank Cass.

Mangan, J. A. (2000) *Athleticism in the Victorian and Edwardian public school: the emergence and consolidation of an educational ideology.* London: Frank Cass.

Mansfield, J. G. (2002) *The Malayan Emergency Campaign from 1945 to 1960: an operations planning process analysis,* Canadian Forces Command and Staff Course 28 (http://www.cfc.forces.gc.ca/papers/csc/csc28/mds/mansfield.doc).

Maslow, A. H. (1959) *Motivation and personality.* New York: HarperCollins.

Massey, D. (1993) 'Power-geometry and progressive sense of place', in J. Bird (ed.), *Mapping the futures: local cultures, global change.* London, New York: Routledge.

Massumi, B. (2002) *Parables for the virtual: movement, affect, sensation.* Durham, NC: Duke University Press.

Massumi, B. (2003) 'Navigating movements: an interview with Brian Massumi' (http://www.21cmagazine.com/issue2/massumi.html).

Massumi, B. (2005a) 'Fear (the spectrum said)', *Positions* 31 (1): 31–48.

Massumi, B. (2005b) The future birth of the affective fact', paper presented at *The sinews of the present – genealogies of biopolitics,* Montreal, May.

Matless, D. (1998) *Landscape and Englishness.* London: Reaktion.

Mbembe, A. (2003) 'Necropolitics', *Public Culture* 15 (1): 11–40.

Mechling, J. (1984) 'Patois and paradox in a Boy-Scout treasure-hunt + folkloristic examination of the role of a game in American culture', *Journal of American Folklore* 97 (383): 24–42.

Mechling, J. (1987) 'Dress right, dress, the Boy-Scout uniform as a folk costume', *Semiotica* 64 (3–4): 319–33.

Mechling, J. (2001) *On my honor: Boy Scouts and the making of American youth* Chicago, London: University of Chicago Press.

Meilinger, P. (1996) 'Trenchard and "morale bombing": the evolution of Royal Air Force doctrine before World War II', *Journal of Military History* 60: 243–70.

Meilinger, P. (1999) 'Air strategy: targeting for effect', *Aerospace Power Journal,* Winter: 48–61.

Meister, D. (1999) *The history of human factors and ergonomics.* Mahwah, NJ, London: Lawrence Erlbaum Associates.

Merriman, P. (2007) *Driving spaces.* Oxford: Wiley-Blackwell.

Milde, M. (2008) *International air law and the ICAO: essential air and space law.* New York: Eleven International Publishing.

Miles, M. E. (1936) *The sea-plane base: an adventure in the Hebrides*. London: A. & C. Black.

Mill, H. (1929) 'The significance of Sir Hubert Wilkins' Antarctic flights', *Geographical Review* 19 (3): 377–86.

Miller, L. and Bloch, R. H. (1984) *Black Saturday: the first day of the Blitz: East London memories of September 7th 1940*. London: THAP Books.

Mindell, D. (2002) *Between human and machine: feedback, computing and control before cybernetics*. Baltimore: Johns Hopkins University Press.

Mirowski, P. (2001) *Machine dreams: economics becomes a cyborg science*. Cambridge: Cambridge University Press.

Moran, Lord (1945) *The anatomy of courage*. London: Robinson.

Morison, F. (1937) *War on great cities*. London: Faber and Faber.

Morrow, J. H. (2003) 'Brave men flying: the Wright Brothers and military aviation in World War One', in R. D. Launius and J. R. D. Bednarek (eds), *Reconsidering a century of flight*. Chapel Hill, NC, London: University of North Carolina Press.

Morse, M. (1990) 'An ontology of everyday distraction: the freeway, the mall, and television', in P. Mellencamp (ed.), *Logics of television: essays in cultural criticism*. Bloomington: Indiana University Press.

Morshed, A. (2002) 'The cultural politics of aerial vision: Le Corbusier in Brazil (1929)', *Journal of Architectural Education* 55 (4): 201–10.

Mort, F. (2004) 'Fantasies of metropolitan life: planning London in the 1940s', *Journal of British Studies* 43: 120–51.

Mortimer, G. (2005) *The longest night: voices from the London Blitz*. London: Weidenfeld & Nicolson.

Muller, B. J. (2004) '(Dis)qualified bodies: securitization, citizenship and "identity management"', *Citizenship Studies* 8 (3): 279–94.

Mutimer, D. (2007) 'Sovereign contradictions: Maher Arar and the indefinite future', in E. Dauphinee and C. Masters (eds), *The logics of biopower and the war on terror: living, dying, surviving*. New York, Basingstoke: Palgrave Macmillan.

Myers, C. S. (1940) *Shell shock in France, 1914–1918: based on a war diary*. Cambridge: Cambridge University Press.

Myerscough, J. (1985) 'Airport provision in the interwar years', *Journal of Contemporary History* 20: 41–70.

Nagl, J. A. (2005) *Learning to eat soup with a knife: counterinsurgency lessons from Malaya and Vietnam*. Chicago, London: University of Chicago Press.

Newman, A., Dixon, G. and Davies, B. (1994) *Towards a study of UK airport retailing: the physical setting and purchasing behaviour*. Manchester: Manchester Metropolitan University, Advanced Institute Research Paper.

Ngai, S. (2004) *Ugly feelings*. Cambridge, Mass.: Harvard University Press.

O'Ballance, E. (1966) *Malaya: the communist insurgent war, 1948–1960*. London: Faber.

Office of United States, Chief of Counsel for Prosecution (1946) *Nazi conspiracy and aggression*, vol. 4. Washington, DC: US GPO.

Ogilvy, G. (1984) *Flying displays*. Shrewsbury: Airlife.

O'Malley, P. (2006) 'Risk, ethics and airport security', *Canadian Journal of Criminology and Criminal Justice* 48 (3): 413–21.

Omar, O. and Kent, A. (2001) 'International airport influences on impulsive shopping: trait and normative approach', *International Journal of Retail and Distribution Management* 29 (4/5): 226–35.

Omissi, D. (1990) *Air power and colonial control: the Royal Air Force 1919–1939.* Manchester: Manchester University Press.

Ong, A. (1999) *Flexible citizenship: the cultural logics of transnationality.* Durham, NC: Duke University Press.

Ong, A. (2006) *Neoliberalism as exception: mutations in citizenship and sovereignty.* Durham, NC, London: Duke University Press.

Orr, J. (2006) *Panic diaries: a genealogy of panic disorder.* Durham, NC: Duke University Press.

Overy, R. (1981) *The air war 1939–1945.* London: Europa.

Paglen, T. and Thompson, A. C. (2006) *Torture taxi: on the trail of the CIA's rendition flights.* Hoboken, NJ: Melville House.

Palmer, R. (1995) 'On wings of courage: public "air-mindedness" and national identity in late imperialist Russia', *Russian Review* 54: 209–26.

Palmer, S. W. (2006) *Dictatorship of the air: aviation culture and the fate of modern Russia.* Cambridge: Cambridge University Press.

Pape, R. A. (1996) *Bombing to win: air power and coercion in war.* Ithaca, NY, London: Cornell University Press.

Paris, M. (1989) 'Air power in imperial defence, 1880–1918', *Journal of Contemporary History* 24: 209–25.

Paris, M. (1993) 'The rise of the airmen – the origins of air force elitism, c.1890– 1918', *Journal of Contemporary History* 28 (1): 123–41.

Paris, M. (1995) *From the Wright Brothers to Top Gun: aviation, nationalism, and popular cinema.* Manchester, New York: Manchester University Press.

Parisi, L. and Goodman, S. (2005) 'The affect of nanoterror', *Culture Machine* (http://culturemachine.tees.ac.uk/Cmach/Backissues/j007/art_res.htm).

Parks, L. (2002) *Cultures in orbit: satellites and the televisual.* Durham, NC: Duke University Press.

Parks, L. (2007) 'Points of departure: the culture of US airport screening', *Journal of Visual Culture* 6 (2): 183–200.

Paschal, H. and Dougherty L. J. (2003) *Defying gravity: contemporary art and flight* Munich, London: Prestel.

Pascoe, D. (2001) *Airspaces.* London: Reaktion.

Pascoe, D. (2003) *Aircraft.* London: Reaktion.

Patterson, I. (2007) *Guernica and total war.* Cambridge, Mass.: Harvard University Press.

Pearman, H. (2004) *Airports: a century of architecture.* London: Laurence King.

Pendo, S. (1985) *Aviation in the cinema.* Metuchen, NJ: Scarecrow Press.

Pevner, T. (2004) 'What a synoptic and artificial view reveals: extreme history and the modernism of W. G. Sebald's realism', *Criticism* 46 (3): 341–61.

Pilger, J. (2006) 'Palestine: a war on children', *New Statesman and Nation*, 15 June (http://www.johnpilger.com/page.asp?partid=401).

Pinder, D. (2004) 'Rivers', in S. Harrison, S. Pile and N. Thrift (eds), *Patterned ground.* London: Reaktion.

Pisano, D. (1993) *To fill the sky with pilots: the civilian pilot training program, 1939– 1949.* Urbana: University of Illinois Press.

Pisano, D. and Van der Linden F. R. (2002) *Charles Lindbergh and the Spirit of St Louis.* Washington, DC: Smithsonian National Air and Space Museum.

Ploszajska, T. (1996) 'Constructing the subject: geographical models in English schools, 1870–1944', *Journal of Historical Geography* 22 (4): 388–98.

Possony, S. and Rosenzweig, L. (1955) 'The geography of the air', *Annals of the American Academy of Political and Social Science* 299: 1–11.

Poszos, R. (2002) 'Nazi hypothermia research: should the data be used?', *Military Medical Ethics* 2: 437–61.

Prince of Wales, Salmond, G., Churchill, W., Amery, C., Haig, E. and Trenchard, H. (1920) 'Imperial air routes: discussion', *The Geographical Journal* 55 (4): 263–70.

Privacy Coalition (2009) 'Stop whole body imaging' (http://privacycoalition.org/stopwholebodyimaging/).

Puar, J. and Rai, A. (2004) 'The remaking of a model minority: perverse projectiles under the specter of (counter)terrorism', *Social Text* 22 (3): 75–104.

Pugilese, J. (2006) 'Asymmetries of terror: visual regimes of racial profiling and the shooting of Jean Charles de Menezes in the context of the war in Iraq', *borderlands e-journal* 5 (http://www.borderlands.net.au/vol5no1_2006/pugliese.htm).

Putney, C. (2001) *Muscular Christianity: manhood and sports in Protestant America, 1880–1920.* Cambridge, Mass., London: Harvard University Press.

Rabinbach, A. (1990) *The human motor: energy, fatigue, and the rise of modernity.* New York: Basic Books.

Raguraman, K. (1997) 'Airlines as instruments for nation building and national identity: case study of Malaysia and Singapore', *Journal of Transport Geography* 5 (4): 239–56.

Rasmussen, N. (2008) *On speed: the many lives of amphetamine.* New York: New York University Press.

Raytheon (2000) 'SIVAM: background and benefits' (http://www.raytheon.com/c3i/c3iproducts/c3isivam/sivam01e.htm).

Reeves, D. M. (1936) 'Aerial photography and archaeology', *American Antiquity* 2 (2): 102–7.

Reid, J. (2006) *The biopolitics of the war on terror: life struggles, liberal modernity and the defence of logistical societies.* London: Palgrave Macmillan.

Reynolds, Q. (1941) *A London diary.* New York: Random House.

Rheinberger, H.-J. (1997) *Toward a history of epistemic things: synthesizing proteins in the test tube.* Stanford, Calif.: Stanford University Press.

Rickman, J. (1938) 'Panic and air-raid precautions: notes for discussion', *The Lancet*, 4 June: 1291–5.

Rieger, B. (2005) *Technology and the culture of modernity in Britain and Germany, 1890–1945.* Cambridge: Cambridge University Press.

Rose, N. (2006) *The politics of life itself: biomedicine, power and subjectivity in the twenty-first century.* Princeton: Princeton University Press.

Rosenthal, M. (1986) *The character factory: Baden-Powell and the origins of the Boy Scout movement.* New York: Pantheon Books.

Rumford, C. (2006) 'Theorizing borders', *European Journal of Social Theory* 9 (2): 155–70.

Said, E. W. (1978) *Orientalism.* London: Penguin.

Saint-Amour, P. K. (2003) 'Modernist reconaissance', *MODERNISM/modernity* 10 (2): 349–80.

Saint-Amour, P. K. (2005) 'Air war prophecy and interwar modernism', *Comparitive Literature Studies* 42 (2): 130–61.

Salter, M. B. (2003) *Rights of passage: the passport in international relations.* Boulder, Colo.: Lynne Rienner Publishers.

Salter, M. B. (2004) 'Passports, mobility, and security: how smart can the border be?', *International Studies Perspectives* 5 (1): 71–91.

Salter, M. B. (2006) 'The global visa regime and the political technologies of the international self: borders, bodies, biopolitics', *Alternatives* 31: 167–89.

Salter, M. B. (2007a) 'Governmentalities of an airport: heterotopia and confession', *International Political Sociology* 1 (1): 49–61.

Salter, M. B. (2007b) 'We are all exiles: implications of the border as state of exception', paper presented at *Contemporary insecurities and the politics of exception*, Standing Group on International Relations Conference, Turin, Italy, September.

Salter, M. B. (2008) *Politics of/at the airport*. Minneapolis: University of Minnesota Press.

Satia, P. (2006) 'The defence of inhumanity: air control and the idea of Arabia', *American Historical Review* 111 (1): 16–51.

Satia, P. (2008) *Spies in Arabia: the Great War and the cultural foundations of Britain's covert empire in the Middle East*. Oxford: Oxford University Press.

Satia, P. (2009) 'From colonial attacks to drones in Pakistan', *New Perspectives Quarterly*, Summer (http://www.digitalnpq.org/archive/2009_summer/08_satia.html).

Saville-Sneath, R. A. (1941) *Aircraft recognition*. London: Penguin.

Scarry, E. (2002) 'Citizenship in emergency: can democracy protect us against terrorism?', *Boston Review: A Political and Literary Forum*, October/November (http://bostonreview.net/BR27.5/scarry.html).

Schivelbusch, W. (1986) *The railway journey: the industrialization of time and space in the 19th century*. Berkeley: University of California Press.

Schivelbusch, W. (2001) *The culture of defeat: on national trauma, mourning, and recovery*. London: Granta.

Schnapp, J. (1994) 'Propeller talk', *MODERNISM/modernity* 1 (3): 153–78.

Schoch-Spana, M. (2004) 'Bioterrorism: US public health and a secular apocalypse', *Anthropology Today* 20 (5): 8–13.

Schwaniger, A. and Hofer F. (2004) 'Evaluation of CBT for increasing threat detection performance in X-ray screening', in K. Morgan and M. J. Spector (eds), *The Internet society: advances in learning, commerce and security*. Southampton: WIT Press.

Schwartz, D. (1998) 'Environmental terrorism: analyzing the concept', *Journal of Peace Research* 35 (4): 483–96.

Scott, J. C. (1998) *Seeing like a state: how certain schemes to improve the human condition have failed*. New Haven, London: Yale University Press.

Sealy, K. R. (1957) *The geography of air transport*. London: Hutchinson University Library.

Seamon, D. (1980) 'Body-subject, time-space routines, and place-ballets', in A. Buttimer and D. Seamon (eds), *The human experience of space and place*. New York: St Martin's Press.

Sebald, W. G. (2003) *On the natural history of destruction*. London: Hamish Hamilton.

Sedgwick, E. K. and Frank, A. (2002) *Touching feeling: affect, pedagogy, performativity*. Durham, NC: Duke University Press.

Sengoopta, C. (2003) *Imprint of the Raj*. London: Macmillan.

Serres, M. (1995) *Angels, a modern myth*. Paris: Flammarion.

Sheller, M. and Urry, J. (2000) 'The city and the car', *International Journal of Urban and Regional Research* 24: 737–57.

Shephard, B. (2001) *A war of nerves: soldiers and psychiatrists in the twentieth century*. Cambridge, Mass,: Harvard University Press.

Sherry, M. (1987) *The rise of American air power*. New Haven: Yale University Press.

Sherwood, J. D. (1996) *Officers in flight suits: the story of American Air Force fighter pilots in the Korean War*. New York: New York University Press.

Sherwood, J. D. (1999) *Fast movers: jet pilots and the Vietnam experience*. New York: Free Press.

Sidaway, J. (2008) 'The dissemination of banal geopolitics: webs of extremism and insecurity', *Antipode* 40 (1): 2–8.

Sidaway, J. (2009) 'Shadows on the path: negotiating geopolitics on an urban section of Britain's South West Coast Path', *Environment and Planning D – Society & Space* 27(6): 1091–116.

Sillitoe, A. (1995) *Life without armour*. London: HarperCollins.

Simmel, G. (1950) 'The metropolis and mental life', in *The sociology of Georg Simmel*, eds D. Weinstein and K. Wolf. New York: Free Press.

Simmons, C. and Caruana, V. (2001) 'Enterprising local government: policy, prestige and Manchester Airport, 1929–82', *Journal of Transport History* 22 (2): 126–46.

Sioh, M. (2004) 'An ecology of postcoloniality: disciplining nature and society in Malaya, 1948–1957', *Journal of Historical Geography* 30: 729–46.

Sloterdijk, P. (2009) *Terror from the air*. New York: Semiotext(e).

Smith, P. and Toulier, B. (2001) *Berlin, Liverpool, Paris: Airport Architecture of the Thirties* Gingko Press, Amsterdam.

Spaight, J. M. (1938) *Air power in the next war*. London: Geoffrey Bles.

Spaight, J. M. (1939) *Can America prevent frightfulness from the air?* London: James Molony.

Sparke, M. (2006) 'A neoliberal nexus: economy, security and the biopolitics of citizenship on the border', *Political Geography* 25: 151–80.

Springhall, J. (1977) *Youth, empire and society: British youth movements, 1883–1940*. London: Croom Helm.

Springhall, J., Fraser, B. and Hoare, M. E. (1983) *Sure & stedfast: a history of the Boys' Brigade, 1883–1983*. London: Collins.

Stein, A. (1919) 'Air photography of ancient sites', *Geographical Journal* 54 (3): 200.

Stewart, D. N. and Winser, D. M. D. (1942) 'Incidence of perforated peptic ulcer: effect of heavy air-raids', *The Lancet*, 28 February: 259–61.

Stoler, A. L. (2008) 'Imperial debris: reflections on ruins and ruination', *Cultural Anthropology* 23 (2): 191–219.

Strachey, J. (1941) *Post D: some experiences of an air raid warden*. London: Gollancz.

Stubbs, R. (1989) *Hearts and minds in guerrilla warfare: the Malayan Emergency, 1948–1960*. Singapore, Oxford: Oxford University Press.

Sulzberger, A. G. and Wald, M. L. (2009) 'Jet flyover frightens New Yorkers', *New York Times*, 27 April.

Sutherland, J. and Canwell, D. (2006) *Battle of Britain 1917: the first heavy bomber raids on England*. Barnsley: Pen & Sword Aviation.

Swann, B. and Aprahamian, F. (eds) (1999) *J. D. Bernal: a life in science and politics*. London: Verso.

Sweet, K. M. (2004) *Aviation and airport security: terrorism and safety concerns*. Upper Saddle River, NJ: Pearson/Prentice Hall.

Tarde, A. (1919) 'The work of France in Morocco', *Geographical Review* 8 (1): 1–30.

Taylor, L. (ed.) (1946) *The story of the Air Training Corps*. London.

Thomas, H. (1920) 'Geographical reconnaissance by aeroplane photography, with special reference to the work done on the Palestine front', *Geographical Journal* 55 (5): 349–70.

Thomas, S. E. (1939) *A.R.P. A concise, fully illustrated and practical guide for the householder and air-raid warden*. St Albans: Donnington Press.

Thompson, L. S. (1955) 'The psychological impact of air power', *Annals of the American Academy of Political and Social Science* 299: 58–66.

Thompson, P. (1992) *The Edwardians: the remaking of British society*. London: Routledge.

Thrift, N. (1996) 'Inhuman geographies: landscapes of speed, light and power', in N. Thrift (ed.), *Spatial formations*. London: Sage.

Thrift, N. (2000a) 'Still life in nearly present time: the object of nature', *Body and Society* 6 (3–4): 34–57.

Thrift, N. (2000b) 'Performing cultures in the new economy', *Annals of the Association of American Geographers* 4: 201–34.

Thrift, N. (2004a) 'Driving in the city', *Theory, Culture & Society* 21: 41–59.

Thrift, N. (2004b) 'Remembering the technological unconscious by foregrounding knowledges of position', *Environment and Planning D – Society & Space* 22 (1): 175–90.

Thrift, N. (2005a) 'But malice aforethought: cities and the natural history of hatred', *Transactions of the Institute of British Geographers* 30 (4): 133–50.

Thrift, N. (2005b) 'From born to made: technology, biology and space', *Transactions of the Institute of British Geographers* 30 (4): 463–76.

Thrift, N. (2006) 'Re-inventing invention: new tendencies in capitalist commodification', *Economy and Society* 35 (2): 279–306.

Thrift, N. and Dewsbury J. D. (2000) 'Dead geographies – and how to make them live', *Environment and Planning D – Society & Space* 18 (4): 411–32.

Ticineto Clough, P., Goldberg, G., Schiff, R., Weeks, A. and Willse, C. (2007) 'Notes towards a theory of affect-itself', *Ephemera* 7 (1): 60–77.

Tomkins, S. S. (1991) *Affect imagery consciousness, vol. 3: The negative effects: anger and fear*. New York, Springer Publishing Co.

Tomkins, S. S. and Demos, E. V. (eds) (1995) *Exploring affect: the selected writings of Silvan S. Tomkins*. Cambridge: Cambridge University Press.

Torpey, J. C. (2000) *The invention of the passport: surveillance, citizenship, and the state*. Cambridge, New York: Cambridge University Press.

Townshend, C. (1986) 'Civilization and "frightfulness": air control in the Middle East between the wars', in C. Wrigley (ed.), *Warfare, diplomacy and politics: essays in honour of A. J. P. Taylor*. London: Hamish Hamilton.

Trotter, W. (1924) *Instincts of the herd in peace and war*. London: T. Fisher Unwin.

TSA (2006) *Recommended security guidelines for airport planning, design and construction*. Transport Security Administration (http://www.tsa.gov).

TSA (2007) 'Innovation: process and technology', Checkpoint Evolution. Transport Security Administration (http://www.tsa.gov/evolution/innovation.shtm).

TSA (2008a) 'Checkpoint evolution: passenger engagement', The TSA Blog, 3 April (http://www.tsa.gov/blog/2008/04/checkpoint-evolution-passenger.html).

TSA (2008b) 'Trollkiller said', The TSA Blog, 31 March (http://www.tsa.gov/blog/2008/03/checkpoint-changes-coming.html).

Tsiamyrtzis, P., Dowdall, J., Shastri, D., Pavlidis, I. T., Frank, M. G. and Ekman, P. (2007) 'Imaging facial physiology for the detection of deceit', *International Journal of Computer Vision* 71 (2): 197–214.

Tymms, F. (1927) *Flying for air survey photography*. Air Ministry. London: HMSO.

Ullman, H. K. and Wade, J. P. (1996) *Shock and Awe: achieving rapid dominance*. Washington, DC: National Defense University Press.

Ullman, H. K. and Wade, J. P. (1998) *Rapid dominance: a force for all seasons; technologies and systems for achieving Shock and Awe – a real revolution in military affairs*. London: Royal United Services Institute for Defence Studies.

Upham, P. (2003) *Towards sustainable aviation*. London: Earthscan Publications.

Urry, J. (2000) *Sociology beyond societies: mobilities for the twenty-first century*. London, New York: Routledge.

Urry, J. (2003) *Global complexity*. Cambridge: Polity.

Urry, J. (2007) *Mobilities*. London: Sage.

Urry, J. (2008) 'Aeromobilities and the global', in S. B. Cwerner, S. Kesselring and J. Urry (eds), *Aeromobilities*. London: Routledge.

Urry, J. and Sheller M. (2006) 'The new mobilities paradigm', *Environment and Planning A* 38 (2): 207–26.

Van der Ploeg, I. (1999) 'The illegal body: "Eurodac" and the politics of biometric identification', *Ethics and Information Technology* 1 (4): 295–302.

Van der Ploeg, I. (2003) 'Biometrics and privacy: a note on the politics of theorizing technology', *Information Communication and Society* 6 (1): 85–104.

Van Riper, A. B. (2004) *Imagining flight: aviation and popular culture*. College Station: Texas A&M University Press.

Vance, N. (1985) *The sinews of the spirit: the ideal of Christian manliness in Victorian literature and religious thought*. Cambridge: Cambridge University Press.

Vaughan-Williams, N. (2008) 'Borders, territory, law', *International Political Sociology* 2 (4): 322–338.

Vaughan-Williams, N. (2010, in press) 'The UK border security continuum: virtual biopolitics and the simulation of the sovereign ban', *Environment and Planning D – Society & Space*.

Verrier, A. (1968) *The bomber offensive*. London: Batsford.

Vidler, A. (2002) 'Photourbanism: planning the city from above and from below', in G. Bridge and S. Watson (eds), *A companion to the city*. Oxford: Wiley-Blackwell.

Vienne, V. (2004) 'Fearmongering: the brand', *AIGA Voice*, February (http://www.aiga.org/content.cfm/fearmongering-the-brand?pff=2).

Virilio, P. (1989) *War and cinema: the logistics of perception*. London: Verso.

Virilio, P. (1997) *Open sky*. New York: Verso.

Virilio, P. (2005) *Negative horizon: an essay in dromoscopy*. London: Continuum.

Vowles, T. M. (2006) 'Geographic perspectives of air transportation', *Professional Geographer* 58 (1): 12–19.

Waddington, C. H. (1973) *Operational research against the U-boat*. London: Elek Science.

Waldau, N. and Schrekenberg, M. (2005) *Pedestrian and evacuation dynamics*. Berlin: Springer.

Waldheim, C. (1999) 'Aerial representation and the recovery of landscape', in J. Corner (ed.), *Recovering landscape: essays in contemporary landscape architecture.* New York: Princeton Architectural Press.

Walford, R. (1969) *Games in geography.* London: Longman Group.

Waller, A. D. (1918) 'Galvanometric records of the emotive response to air raids', *The Lancet*, 23 February: 311.

Walmsley, L. (1920) 'The recent trans-African flight and its lesson', *Geographical Review* 9 (3): 149–60.

Walters, W. (2006) 'Border/control', *European Journal of Social Theory* 9 (2): 187–203.

Warden, J. (1995) 'Air theory for the twenty-first century', in Air Chronicles (ed.), *Battlefield of the future: 21st century issues* (http://www.airpower.maxwell.af.mil/air-chronicles/battle/chp4.html).

Weber, C. (2007) 'Securitizing the unconscious: the Bush doctrine of preemption and *Minority Report*', in E. Dauphinee and C. Masters (eds), *The logics of biopower and the war on terror: living, dying, surviving.* New York, Basingstoke: Palgrave Macmillan.

Weizman, E. (2002) 'The politics of verticality', *Open Democracy*, 1 May (http://www.opendemocracy.net/conflict-politicsverticality/article_810.jsp)

Weizman, E. (2007) *Hollow land: Israel's architecture of occupation.* London: Verso.

Wells, H. G. (1908) *War in the air* (http://ebooks.gutenberg.us/WorldeBookLibrary.com/warair.htm).

Wells, M. K. (1995) *Courage and air warfare: the Allied aircrew experience in the Second World War.* London: Frank Cass.

Westerman, J. F. C. (1938) *Menace from the air.* London: Oxford University Press.

Westing, A. (1983) 'The environmental aftermath of warfare in Vietnam', *Natural Resources Journal* 23 (April): 365–89.

Whitaker, O. (1917) 'Aeronautical charts', *Geographical Review*, July: 3–4.

Whitehead, M. (2009) *State, science and the skies: governmentalities of the British atmosphere.* Oxford: Wiley-Blackwell.

Whitelegg, D. (2005) '"Places and spaces I've been": geographies of female flight attendants in the United States', *Gender, Place and Culture* 2: 251–66.

Wiener, N. and Bigelow, J. (1943) 'Behaviour, purpose and teleology', *Philosophy of Science* 10 (1): 18–24.

Wijninga, P. W. W. and Szafranski, R. (2000) 'Beyond utility targeting toward axiological air operations', *Aerospace Power Journal*, Winter: 45–59 (http://www.air-power.maxwell.af.mil/airchronicles/apj/apj00/win00/szafranski.htm).

Wilds, D. (2006) 'Anthropometry', *Flight Fax: Army Aviation Composite Risk Management Information* 34 (5): 8–9.

Wilkins, H. (1929) 'The Wilkins–Hearst Antarctic expedition, 1928–1929', *Geographical Review* 19 (3): 353–76.

Wilkinson, P. and Jenkins, B. M. (1999) *Aviation terrorism and security.* London: Frank Cass.

Williams, A. (2007) '*Hakumat al tayarrat*: the role of air power in the enforcement of Iraq's boundaries', *Geopolitics* 12 (3): 505–28.

Williams, A. (2009) 'A crisis in aerial sovereignty? Considering the implications of recent military violations of national airspace', *Area* 3: 505–28.

Wilson, D. and Weber, L. (2008) 'Surveillance, risk and preemption on the Australian border', *Surveillance and Society* 5 (2): 124–41.

Wilson, H. (1942) 'Mental reactions to air-raids', *The Lancet*, 7 March: 284–7.

Wohl, R. (1994) *A passion for wings: aviation and the Western imagination, 1908–1918*. New Haven: Yale University Press.

Wohl, R. (2005) *The spectacle of flight: aviation and the Western imagination, 1920–1950*. New Haven, London: Yale University Press.

Wood, A. (2003) 'A rhetoric of ubiquity: terminal space as omnitopia', *Communication Theory* 13 (3): 324–44.

Woodhouse, H. (1917) 'Aeronautical maps and aerial transportation', *Geographical Review* 4 (5): 329–50.

Woodward, R. (2004) *Military geographies*. Oxford: Blackwell.

Wright, J. J. (1952) *Geography in the making: The American Geographical Society 1851–1951*. New York: The American Geographical Society.

Wrigley, C. (2002) *Winston Churchill: a biographical companion*. London: ABC.

Zeleny, J. (2009) 'Aide quits over flight that alarmed New York', *New York Times*, 8 May.

Zenko, M. (2009) 'Say no to a Darfur no-fly zone', *The Guardian*, 12 March (http://www.guardian.co.uk/commentisfree/cifamerica/2009/mar/12/darfur-no-fly-zone).

Zook, M. A. and Brunn, S. D. (2006) 'From podes to antipodes: positionalities and global airline geographies', *Annals – Association of American Geographers* 96 (3): 471–90.

Zuckerman, S. (1932) *The social life of monkeys and apes*. London: Harcourt, Brace and Company.

Zuckerman, S. (1940) 'Experimental study of blast injuries to the lungs', *The Lancet*, 24 August: 219–24.

Zuckerman, S. (1941) 'The problem of blast injuries', *Proceedings of the Royal Society of Medicine* 34: 171–88.

Zuckerman, S. (1978) *From apes to warlords: the autobiography of Solly Zuckerman*. London: Hamilton.

Index